Praise for *Wooden Boats* by Michael Ruhlman

"[Ruhlman] manages to tell the story of the Martha's Vineyard boatyard, Gannon and Benjamin Marine Railway . . . with eloquence. Although Ruhlman comes to the project with little boating experience . . . he shows he not only understands and values the process intrinsically, but also that he is a writer with the talent to bring the reader along with him on the job . . . we begin to understand the extraordinary blend of creativity, engineering, and craftsmanship called for in the building of traditional vessels in modern times."   —*WoodenBoat*

"With the same storytelling skills he brought to *The Making of a Chef* and *The Soul of a Chef*, Ruhlman captures the spirit of the craftsmen who work at a demanding trade and produce an artwork of great utility and beauty. This narrative will be savored by armchair sailors, admirers of painstaking craftsmen, and those who nurture a passion for the sea and all pertaining to it."   —*Richmond Times-Dispatch*

"That passion [for wooden boats] is evident in every word of Ruhlman's account. A love of wooden boats will attract many readers to this book, but those who know little about wooden boats will find a readable entrée into this romantic world."   —*Cape Cod Times*

"[Ruhlman's] ability to simply tell the boat-builders' story, making connections between boats and life, gives this sharply observed book its pleasures."   —*Publishers Weekly*

PENGUIN BOOKS

WOODEN BOATS

Michael Ruhlman has written for *The New York Times,* the *Los Angeles Times, Gourmet,* and *Food Arts* magazine and is the recipient of a James Beard award for magazine writing. He is the author of *Boys Themselves, The Making of a Chef,* and *The Soul of a Chef* (available from Penguin), a *New York Times* Notable Book. He lives in Cleveland Heights, Ohio, with his wife and children.

Wooden Boats

In Pursuit of the Perfect Craft
at an American Boatyard

MICHAEL RUHLMAN

PENGUIN BOOKS

PENGUIN BOOKS

Published by the Penguin Group

Penguin Putnam Inc., 375 Hudson Street, New York, New York 10014, U.S.A.
Penguin Books Ltd, 80 Strand, London WC2R 0RL, England
Penguin Books Australia Ltd, 250 Camberwell Road,
Camberwell, Victoria 3124, Australia
Penguin Books Canada Ltd, 10 Alcorn Avenue,
Toronto, Ontario, Canada M4V 3B2
Penguin Books India (P) Ltd, 11 Community Centre,
Panchsheel Park, New Delhi – 110 017, India
Penguin Books (N.Z.) Ltd, Cnr Rosedale and Airborne Roads,
Albany, Auckland, New Zealand
Penguin Books (South Africa) (Pty) Ltd, 24 Sturdee Avenue,
Rosebank, Johannesburg 2196, South Africa

Penguin Books Ltd, Registered Offices:
Harmondsworth, Middlesex, England

First published in the United States of America by Viking Penguin,
a member of Penguin Putnam Inc. 2001
Published in Penguin Books 2002

1   3   5   7   9   10   8   6   4   2

THE LIBRARY OF CONGRESS HAS CATALOGED THE HARDCOVER EDITION AS FOLLOWS:
Ruhlman, Michael, 1963–
Wooden boats / Michael Ruhlman.
p.   cm.
ISBN 0-670-88812-5 (hc.)
ISBN 0 14 20.0121 X (pbk.)
1. Rebecca (Wooden yacht)   2. Yacht building—Massachusetts—Vineyard Haven.
3. Ruhlman, Michael, 1963–   4. Benjamin, Nat.   5. Gannon, Ross.   6. Schooners—
Design and construction.   7. Ships, Wooden—Design and construction.   I. Title.
VM331   .R94   2001
623.8'2023—dc21   00-047738

Printed in the United States of America
Set in Bembo
Designed by Jaye Zimet

For Donna

Contents

Prologue

"If you're going to write about wooden boats," Jon Wilson said, "you've got to at least *talk* to Nat and Ross."

I'd come to Maine looking for a boatyard. Maine was the country's wooden boat center, with more wooden boat yards by far than any other state. Traditional wooden sailboats and workboats still plied these waters in significant numbers relative to the rest of the country, which had gone completely fiberglass decades ago. They were still being made here the old-fashioned way—by craftsmen at small yards. Maine, with its craggy coast and clear, cold waters, was the *spiritual* center of wooden boats as well. These boats were as much a part of its romance and character as the evergreen forests and granite outcroppings that marked its shore. It was here that Jon Wilson had founded his magazine, *WoodenBoat,* whose offices were set up in an old white mansion overlooking the blue waters of the Eggemoggin Reach in the tiny hamlet of Brooklin. I explained to Jon that I wanted to work at a wooden boat yard to learn about what I had reason to suspect was an unusual world, and that Maine was the place to do it, not tony Martha's Vineyard, where these Nat and Ross fellows happened to be.

"Nat and Ross," Wilson replied, "are doing in Vineyard Haven what everyone thinks is happening in Maine but isn't." Nat Benjamin, he went on, was one of the best designers and purest builders in the country, an artist, articulate and vocal about the importance of traditional wooden boats, plank-on-frame boats. Benjamin was crazy for the gaff rig, a type of sail plan that had been out of fashion since the 1930s. And finally, Wilson said, Benjamin and his partner, Ross Gannon, were real sailors; they knew firsthand the kinds of pressures the sea put on a boat and built their boats according to that, rather than current notions of contemporary design. And they had begun an extraordinary new boat, one of the most exciting constructions going on in the country, a 60-foot schooner named *Rebecca*.

In my quest for a boatyard, I'd asked to meet with Wilson because he was, of course, more than just the founding publisher of *WoodenBoat* magazine, a glossy bimonthly filled with gorgeous photographs of and stories about all forms of traditional and contemporary wooden boats. He was The Man, the central figure in the contemporary wooden boat universe, the visionary who, nearly twenty-five years earlier, had seen this universe about to be extinguished by yacht construction's all but absolute embrace of fiberglass and plastics—and against all odds, and *sense,* had published a magazine devoted to the wooden boats that were no longer being built. It had succeeded beyond all logic.

Within a decade, *WoodenBoat* circulation had surpassed 110,000, more than ten times the number of yards and wooden boats then in existence. While this speck of an industry generated a circulation far disproportionate to its numbers, the biggest mass-market magazine, *Boating,* devoted to the entire yachting world—a milieu in which thousands of boatyards pumped out millions of boats in America—had a circulation of only 202,000. Somehow Wilson had captured people's imaginations. People who didn't even go out on the water bought his magazine. He knew all along that the subject of his focus, the wooden boat, would prove to be far more important than its dwindling numbers suggested.

By the time I arrived in Wilson's office in the fall of 1998, the industry was no longer in danger of dying out, and Wilson was considered its savior. Crowds at wooden boat shows and launchings parted like the Red Sea when he walked toward a new boat. His words about it would be accepted as the benediction, the crowd silent and still so as not to miss a single note or nuance of the galvanizing elixir, Jon Wilson's calm, sure voice.

Friends and associates described a more subdued version of the man. One called him the consummate salesman of ideas. Another said that no idea uttered in Wilson's presence went unexplored: no matter how banal it might seem on the surface, Wilson had a way of getting inside other people's ideas and pushing them out to their most provocative boundaries.

Wilson was less than imposing in person, dressed in jeans, standing about five foot eight, trim, with wavy, dark hair well this side of unruly. He appeared light and filled with energy, taking the carpeted stairs of this lush house two at a time toward his office, past Xerox machines wedged into hallways, stopping to say hello to Matt Murphy, the thirty-four-year-old editor of *WoodenBoat* and Wilson's replacement four years earlier. Wilson strode into his office, a cluttered former master bedroom of the old house, and right behind him was Kim Ridley, the editor of his latest venture, *Hope,* a magazine attempting to offer solutions to disturbing social issues. Kim had images from El Salvador and black-and-white photographs of a bare torso, exquisite self-portraits of a woman who had undergone a double mastectomy. Downstairs in the living room—a discreet store selling a variety of books and tapes and paraphernalia pertaining to wooden boats—a steady stream of worshipers came to browse the materials on the shelves and examine the half models on the walls, to think about wooden boats, to see the place where it all happened and maybe even catch a glimpse of The Man himself.

Down the road on this estate was a redbrick barn that had been turned into a school, founded by Wilson, where wooden boat building was taught, advancing the knowledge of wooden boat construction and increasing the numbers of those who could actually practice the craft. At the shore, a big, comfortable boathouse overlooked a harbor filled with wooden boats, the center of the school's summer sailing lessons, designed to further knowledge of basic sailing and general seamanship. All of it was the result of Wilson and the success of his magazine.

In the world of wooden boats, Jon Wilson was the spiritual leader, the holy man. He was also a big part of the reason each new big boat and the one before it and the one after it were able to be built. The magazine had created a collective voice of boatbuilders and boat lovers and sailors and designers and buyers who before had been scattered and largely cut off from one another in the new world of fiberglass and hulls popped out of molds. It was Wilson who had given a dying industry a clear and beautiful song, sung by a newly unified chorus.

Such was the aura that surrounded Jon Wilson. Hear him! He would announce it not with fear or embarrassment but as fact: *"Wooden boats . . . are alive,"* he said. *"Wooden boats . . . have a soul."*

All wooden boats were beautiful, he proclaimed, "as if the grace of the forest trees were bequeathed in abundance to every plank sawn."

He basked in the maintenance a wooden boat demanded: "The care of living things requires deeper commitment and responsibility."

And he declared, "I truly believe that wooden boats have a lot to teach us about our purpose on the planet."

Our purpose on the planet! Jon Wilson dwelled in this order of magnitude.

Wilson denigrated his own story. All it amounted to, he said, was "how a little nobody got to be editor of *WoodenBoat*." But when he said that anyone looking to write about wooden boats had to at least talk to Nat Benjamin and Ross Gannon, you listened.

Having traveled with my family from Cleveland, our home, to Maine only to be told that I needed instead to be somewhere off Cape Cod was disheartening until Wilson's brow knitted, and he paused. "Ya know," he said, "I think *When and If* is going to be here this week. Nat and Ross might be here."

Divine coincidence? No, a boat launching down in Rockport, an hour south of Brooklin, the reason I happened to be here this particular week as well. A 76-foot racing sloop was to be lowered into Rockport Harbor in a few days, and people came out for such an event. It was like a birth in the community, a champagne christening and the celebration of new life.

✳

This was all new to me. I was not a sailor or a boatbuilder, I had no particular affinity for boats generally, and I had no carpentry skills. I had only recently begun to read about wooden boats after a colleague suggested that I write about them. This colleague, a wooden boat owner himself, had done more than just suggest wooden boats as a possible subject, however. He had during an inspired half hour painted an amazing picture of wooden vessels, which were nearly as old as mankind itself and still vital today. Wooden boats combined extraordinary craftsmanship with centuries of wisdom about how to keep pieces of wood together at sea—*pieces of wood,* planks bent over frames and fastened with bronze. The most basic questions were intriguing. How did they work, how did you keep the water out? Science, the physics of it, was involved in a way that was artful. These boats were almost invariably beautiful to behold. The science and beauty were inextricably linked, were perhaps the same thing. They were instruments, this colleague claimed, as finely rendered as a Stradivarius but strong enough to withstand gale winds and crashing waves over decades, and they cruised through a culture that was filling up with homely plastic objects that didn't last. This was what Wilson had tapped into; this was why his magazine had become disproportionately successful. The wooden boat was a metaphor for all the things that mattered in our cheap, disposable culture.

My colleague hooked me ultimately when he moved on to a description of the people, these artisan builders. Rarely was a working class so enmeshed with an upper class, the wealthy and well heeled who paid for their product, as in the world of wooden boats. In few places anywhere did the rich and successful and famous revere the working class more than in this world. At classic regattas, the virile billionaire vied for the boatwright's attention. Here the wooden boat builder was the Brahmin. Moreover, this was a world filled with sailors of all stripes—I knew there were great stories to be told here.

And so this wooden boat lover convinced me to focus my landbound gaze and spirit on the wooden boat and its builder. I found the idea of it immediately appealing. I knew he was right while he spoke, I needed no reflection: the veracity of what he said, the reason for his personal love of wooden boats, were self-evident and only deepened as I pursued them.

And so I set out to find a wooden boat yard and go to work there, the only way to understand an unfamiliar culture, a new world: through its work. Through their work are people known; through work a new language is learned. The world of wooden boats was well known to itself, of course, like any insular fraternity, but it was not widely written about or understood. I wanted to paint this world, and there wasn't a blanker canvas than my own.

What was this wooden boat, really? Who were the people who built it, and who had they become because of this work? Why did it have such a hold on so many people's imaginations? How did you shape those planks? How did you keep the water out?

And ultimately, was Jon Wilson right about his grandest claim? He was no idiot, that was for sure, but it was entirely possible that he was simply a deluded romantic who had happened to convince a hundred thousand or so people of his vision—the Reverend Moon of wooden boats. But that wasn't likely, either. What if he was right, what if he was even *half* right, in his preposterous but deeply held conviction that wooden boats had something to teach us about our place on the planet? If *that* was true, I wanted to know.

That night, Wilson called me at the house up the road in Blue Hill where my family and I were staying. He told me that Nat and Ross, presumably two of the country's best builders of these objects, were in fact expected in Rockport Harbor on the schooner *When and If* for the new boat's launching, and that he'd be willing to introduce me if I wished.

✳

The crowd appearing for the launch at Rockport Marine that clear, cold September morning numbered in the hundreds, but I suspected immediately that the man in the blue work pants and untucked flannel shirt was either Nat Benjamin or Ross Gannon. *Hoped,* rather—if he was, then I sensed that everything Wilson had told me would prove correct. This man had thick, wavy, sun-blond hair and a brown, graying beard, and against a crowd decked out in the colors of Hilfiger and Patagonia, he seemed to have stepped straight out of the Maine woods. I watched him nod, vaguely skeptical, as Donald Tofias, a Massachusetts real estate developer who had commissioned the 76-footer that was now suspended in the Travelift and about to be christened *White Wings,* made a speech. Jon Wilson, in faded jeans and shades, watched from the crowd, but central, as always. *White Wings* was lowered into the high tide, and when the fanfare subsided, Wilson introduced me to Ross Gannon. Ross smiled from within that hair and beard—his eyes a shocking blue, as if they still carried the reflection of the water—and said he'd be happy to talk to me about wooden boats.

We walked back to *When and If,* a husky, black-hulled schooner, and sat below deck. One of the main ideas Gannon talked about was how durable wooden boats were compared with fiberglass and plastic, how you could always fix them—they could last forever if you took care of them, and unlike with fiberglass boats, you *wanted* them to last forever. *When and If,* originally built for General George S. Patton in 1939, had smashed on rocks during a storm in 1990; Ross and his crew at Gannon & Benjamin Marine Railway had rebuilt her port side and reconfigured the cabins below deck, and here she sat, at rest in the harbor, gawkers loitering on the dock just staring at the creature. I quickly saw an intensity that belied Gannon's laconic, easy manner. My three-year-old daughter was up and down the companionway, frolicking through the spacious quarters of the 63-foot boat, enchanted as children invariably are to be down below in a boat. And she was in view in the main salon when I asked Gannon why wooden boats were important to him—why had he devoted his life to them? Ross seemed surprised by my apparent ignorance regarding what to him was plain, and his blazing eyes burned right through me.

"Do you want to teach your daughter that what *you* do, what you care about, is disposable?" he asked. "That you can throw your *work* away? It doesn't *matter?*"

Gannon's partner, Nat Benjamin, had arrived on the dock with his wife, Pam, and was encircled by a bevy of friends as I stepped off. Nat was laughing. He wore a snappy vest and chinos and had a short beard and long, reddish hair that curled almost to his shoulders. He was fifty-one, like his partner. I spoke with Benjamin for a few minutes, but I already knew what I wanted, having by that time visited several shops in Maine. I told Nat Benjamin I'd like to visit him and his shop. He said he'd be sailing back to Martha's Vineyard in a few days and that I was welcome anytime.

I thanked Ross Gannon, too, and said I hoped to see their shop soon. He grinned and said, "I'll meet you at the ferry."

✳

I doubt that either Nat or Ross thought much about me till I really *did* show up, to watch the construction of this great schooner *Rebecca* and a second boat, a 32-foot powerboat modeled after a 1930s work-boat, and to meet and know the people who did this work, here, on what I would discover was an unusual spot of beach, the waterfront of Vineyard Haven. Gannon didn't meet me at the ferry (when I arrived, a month later, he was back in Maine, picking up a load of wood and tomcatting a lithe fisherman named Kirsten whom he'd become re-acquainted with that day back in Rockport), but both he and Nat Ben-jamin were as good as their word. They welcomed me in the best possible way: they put me to work.

One

Rebecca

I

oatstruck: there could be no other explanation for the impulse toward *Rebecca*. The man was boatstruck. Some people become boat smart; others are simply struck. Something happens to certain men when they see a boat, and they become crazy. A man, or the occasional woman (women seem to be less frequently disturbed), who is boatstruck shows no easily discernible outward signs of the illness. On the contrary, the boatstruck look more than reasonable. They are successful people. They are not easily carried away. They have accumulated if not substantial wealth, then at least significant disposable income. They are smart, cool, self-possessed, and they are pretty good on the water. They brim with a free and adventurous spirit. You tend to like these people—they can be inexplicably magnetic. But a man who is boatstruck often has an unrealistic understanding of his cash situation. And cash is the fulcrum on which a boatstruck life teeters between bliss and ruin. Boats require plenty of cash.

And yet there is something exquisite about the condition of being boatstruck. An ecstasy runs through it, compulsive and contagious. You can see it, sense this delight, even if you happen to be free of the affliction yourself or don't sail or even if you don't particularly care for boats. Sometimes a beautiful boat is simply worthy of devotion, reverence, and awe, and no one doubts it. A beautiful boat is as obviously invaluable as a Leonardo sketch or Monet's water lilies. The boat can be a magnificent structure. And the boat most likely to be deemed so is surely the wooden boat. Many groan at the thought of such boats, recalling some youthful foolishness that resulted in much maintenance and repair and not a single second of actual sailing. That's a wooden boat, all right, but that boat is not magnificent. We are not talking here about the wreck, that piece of shit on the farthest mooring, built in the fifties and uncared for almost from its launching; we are talking rather about the well-built, meticulously crafted, lovingly cared for, continually sailed, plank-on-frame, gaff-rigged vessel. *That* boat inspires. It can be 20 feet or 80, and it is the same thing—it's a form that you know at your core. The perfect wooden sailboat. Such a product of man's mind and labor—a series of pieces of wood bent around frames—is worthy of even the most peculiar fetish of man, of adoration, is worthy perhaps even of lunacy.

In 1995, in Vineyard Haven, Massachusetts, a boatstruck man answered a question posed by a boatbuilder. It may have been one of the most common of all dialogues between the boatstruck and the boatbuilder, but on this occasion something substantial *would* be born from it, born of all this, all that wooden boats were and are, all that they attract—this adoration, this reverence, this innate sense of truth, this *want,* this insanity, this intelligence, this capacity to imagine beauty and draw its design so that it will move through water with grace and power, be drawn through it by wind—a boat born ultimately of a deep knowledge of how wood on water works, knowledge earned over many thousands of miles on the earth's oceans, many decades building boats, and long study of a five-thousand-year-old practice, and born, too, of a sense of all that the boat might be, simply for the boat: *what might be.* Out of all this came *Rebecca.* It floated out of a collective unconscious like a ghost ship materializing out of a fog.

It was to be the last great plank-on-frame schooner built in the twentieth century: *Rebecca.* This wooden yacht may have been sparked by a particular form of ecstatic insanity—that of a man boatstruck—but the thing itself was as solid and durable as the hardwoods that would compose its backbone, a deep-water sailing vessel powerful enough to cruise any of the earth's oceans, a beautiful boat, a mighty ship, one that would exalt all who beheld her. If she could be built.

✴

Rebecca's precarious history began, ironically, in Fort Lauderdale, Florida—one of the plastic-boat meccas of the world, with its blacktop, heat, traffic, and fiberglass—where a boat called *Jane Dore III* was tied up. She had been purchased by a man in Massachusetts. The seller in Fort Lauderdale had called a friend, one Ross Gannon, and asked him to deliver *Jane Dore III,* sail her up to her new owner, who lived on Cape Cod. It was to be a routine delivery for a routine owner.

Ross Gannon could hardly have imagined the course of events that he was setting in motion when he, Suzy Zell, their son, Lyle, and two other crew flew down to Fort Lauderdale to sail the 53-foot center-cockpit yawl to Vineyard Haven Harbor on behalf of the buyer, Daniel Adams. Boat deliveries are always dicey—you never know what you're in for, particularly when it comes to wooden boats, which, when they're being sold, are invariably old and tired and leak like hell. *That's why they're being sold!* No one sells a beautiful wooden boat in excellent condition that's great to sail—he'd be a fool. Boats like that, people keep: that's why they have them in the first place. What happens is that someone neglects a great wooden boat for many years, and when it gets to be too expensive and too much of a headache to repair, *then* he sells it. He simply waits for a boatstruck man to come along, and watches as the man signs over a large check without so much as a ma-rine survey, just scratching his name across the check.

A man who is boatstruck does not realize that good used wooden boats are not available, nor can he differentiate between the gorgeous fantasy in his head and the floating piece of rot in front of him: he sim-ply can't see it. Which helps to explain why Dan Adams would have bought a boat without a marine survey—a matter of paying someone knowledgeable a small fee to look the boat over for any significant problems. No, Dan simply saw the boat and wrote out a check for half the price right there, and in May 1995 four crew led by Ross Gannon flew to Fort Lauderdale, spent a week readying the boat for its voyage north, and then set sail.

If you had to deliver a tired wooden sailboat, one that hadn't been surveyed, up the entire East Coast of the United States to Massachu-setts, you wouldn't be unhappy to have Ross Gannon aboard. He is an able captain and an ingenious mechanic, and he knows how to get you safely where you need to go. At the time of the delivery, in 1995, Ross was forty-eight; he was strong as a draft horse from his work and had decades of experience in deepwater sailing. Moreover, he could fix absolutely anything—the engine, a hull leak, broken wood, a ripped sail, snapped rigging, a busted tiller. If the electronic GPS (global positioning system) went on the fritz, he could surely fix it, too, but he probably wouldn't, since he doesn't like to rely on electronic gadgets to tell him where he is; they fail too often to be relied on. Eighteenth-century navigation techniques work fine, are more reliable, and are more interesting besides. Before he and his partner started their boatyard, the Gannon & Benjamin Marine Railway, Ross tore down old, unwanted houses on the Vineyard and used the material to build beautiful brand-new ones. He was and is an amazing scavenger. If he needs an expensive tool, he usually finds a broken version that someone has thrown away and fixes it according to the job he needs it for. Sometimes he simply builds the tool himself with a blowtorch and scraps from around the yard. He used to move whole houses hundreds of feet when beach erosion threatened them. It was the work of a morning for Ross to jack up an entire house so he could put in a foundation. Ask anybody at the boatyard: there isn't anything Ross can't do when it comes to fixing, building, or moving, not anything.

Three weeks after leaving Fort Lauderdale, *Jane Dore III* reached Vineyard Haven Harbor, its crew safe but sodden (the decks leaked like the devil; foul-weather gear was eventually renamed *Jane Dore* pajamas). The wind was out of the north, blowing straight in the face of the harbor, and *Jane Dore III* came screeching toward the town dock. Ross waved to Nat as he passed *Venture,* Nat's 37-foot sloop, on her mooring near the breakwater, and Nat smiled and waved back, always happy to see another wooden boat arrive in this harbor that was filling up with wooden boats, their masts of varnished spruce glowing in the sun. As Ross approached the town dock, he was forced by the windy conditions and the speed of the boat to throw a line to a stranger for assistance—a big fella in a black leather jacket—and hope that the man would know what to do. He called to him, "Can you tie a bowline?" A bowline, of course, is one of the first sailing knots everyone learns as a kid and may be the most often used knot on a boat. For a sailor, it's like tying his or her shoe. Ross just wanted someone to tie a bowline and throw the line over a piling till he could attend to it.

The man caught the rope and said gravely, *"I'm Dan Adams. I can tie a bowline."* (Ross smiles today at the recollection, saying, "And I think he did. Maybe on the first try!") The guy tying the bowline was the new owner, the boatstruck man.

The inevitable was soon discovered: *Jane Dore III* was held together by paint, Bondo, duct tape, and little else. Nat and Ross looked the boat over and gave Dan the news: it would cost about $250,000 for a complete restoration, anything short of which would be a waste of money. But they didn't recommend a complete restoration. For the same price, but probably less, they could get the designs, salvage the ballast, engine, spars, and hardware, and build Dan a new boat. Nat and Ross tried to console Dan by saying, *Yes, it'll be expensive, but look at it this way: when we're done, you're gonna have the boat you've always wanted.*

After a troubled moment or two, it occurred to Dan that they had a point. Dan Adams had first been boatstruck at the age of five, when his grandfather took him to a dock in Hyannis, on Cape Cod, where a 65-foot Alden was tied up. Since the moment he'd laid eyes on that boat, he said, he had wanted a 65-foot schooner. He would tell you he'd grown up in Boston, attended prestigious Roxbury Latin, an independent boys' school, and studied briefly at the University of Vermont before dropping out to work in politics and, later, movies. But he'd been boatstruck at age five, and after thirty-three years, the condition, untreated, had evidently become quite serious. And so when he heard those words—*you're gonna have the boat you've always wanted*—Dan stopped, thought, and said, *That's not the boat I've always wanted.*

Nat Benjamin's deep voice and set teeth combine in a charismatic grin and laughter so natural that he can charm the president, senators, and movie stars as easily as he does the boatyard workers. It's the charm of authenticity—he's a true sailor, a talented builder and craftsman, and you sense it the moment you lay eyes on him. Here is a man to contend with, here is a man you want on your side. What comes with this authenticity is the ability to be terribly convincing when he talks about boats. So when Dan made his statement *That's not the boat I've always wanted,* Nat asked, *What is?*

Then Dan uttered a seemingly benign but in fact dangerous, life-altering phrase. He said, "A sixty-five-foot schooner."

"We perked up at that!" Nat would later exclaim, laughing his low, easy laugh, unconcerned at the time that this might be just a little too easy; Dan did show every outward sign of being able to afford this. "Very sensible decision!"

And thus was the seed planted in the boatstruck Dan. Nat and Ross could *build* him the very boat that had been in his mind since he was a boy. She could now be his, and he'd be able to sail her on any ocean on this earth, an amazing plank-on-frame schooner. She was to be his.

✳

This was a heady commission for Nat Benjamin. As those involved with boats, and certainly all those involved in designing and building boats, will tell you, they're continually thinking about their ideal boat, what it would be, how they would do it. And now, for Nat, an actual commission for exactly this had walked right onto his and Ross's dock.

Nat's favorite boats were always in the 60- to 70-foot range. He now owned, with Ross and another partner, the 63-foot schooner *When and If.* He and Ross had previously owned a 72-foot yawl named *Zorra,* which they'd chartered yearly in the Caribbean and Martha's Vineyard. That was the perfect cruising and chartering size, Nat thought. Get much bigger and you needed more manpower and sailing experience aboard to handle the boat properly; smaller than that and you began to limit the number of people who could cruise comfortably for extended periods. And you could sail a boat that size anywhere in the world. Dan Adams didn't really need the 65 feet he'd asked for, Nat decided; 60 would do just fine. Nat didn't think Dan had any clear idea of what he really wanted, anyway; he could see the flickery eye and hear the stumbling, vague speech that marked a boatstruck man, and so he himself would decide what would be best for the *boat,* and therefore for Dan.

More important than the ideal length was the kind of rig Dan wanted: a schooner rig, with two masts, like *When and If,* as opposed to a sloop (with one mast) or a ketch (two masts, the after mast the shorter of the two) or a yawl (a big mainmast and a tiny mast way back aft, like *Zorra*).

"*When and If,*" Nat told me, moving in his mind through the schooners he'd known. "Going back many years ago, sailing *Malabar X* in the Caribbean, a fifty-eight-foot gaff-rigged John Alden schooner—sailed her once, but it was a memorable thing. And I sailed a little Block Island schooner across the Atlantic. I sailed a great deal with a friend in the West Indies, on a forty-foot Tancook schooner—simple, clean, nice sailing boat. The first charter I ever did, I crewed on a schooner called *Madrigal* out of Essex, Connecticut, a really nice Alden schooner. I was nineteen years old. That was the first one I sailed on, and it was just a wonderful boat.

"So I've always loved the rig. I love how they sail, I love how easy they are to handle. If you take [two boats with] the same hull and you put a sloop rig on one and a schooner rig on the other, the sloop at first seems simpler. A schooner, gosh, you got two jibs, a foresail and a mainsail, and you may be able to put a fisherman or topsail on top of this. But what makes the schooner simpler is that you've got the same amount of sail area to drive this hull, but you can divide it up into four sails, so each sail is so much easier to handle. It's easier on the boat, and it's a much more versatile rig. Leaving a mooring with a sloop, as soon as you put up a mainsail, that boat is trying to go. The mainsail on a schooner, it's farther aft, so you set that, it acts like a weather vane—it doesn't make the boat go, it makes it stand still, right into the wind. If you're anchoring, coming to the anchor, drop your headsails, jog it in, you ease the fore off and strap the main in, she's gonna sit there like a well-trained horse. So even though there are a lot more strings to pull, each sail is smaller, easier to manage.

"There are more things to do. In a sloop, you set the main and the jib up and that's it. That's what you get, and that's OK, but on a schooner, if the wind starts to blow like hell, you can drop the foresail and the jib and go under just the main and the forestaysail. Or you can drop the main and go under fore and forestaysail—there's all sorts of options. And that makes it fun."

When Nat designed a boat, he thought not only about where and how her future owner would sail her but also about how *he* might sail her, what he'd like to do if he was aboard. "It's fun having a bowsprit," he said, "because it's a great place to sit—you just go out there, climb out, you're away from everyone and you can climb down and put your feet in the water, or you can go up the rig and sit on the spreaders. That's what I like about these boats; you don't just have to sit in the cockpit and talk to people, you can say, 'Excuse me, I gotta go check something,' and, boom, you're up the mast and in another world, and that to me means a lot.

"I just think schooners are more fun to sail. And prettier to look at."

So on that day in 1995, Nat walked the mile and a half from the boatyard to his home on Grove Street—where he and his wife, Pam, have lived since the mid-1970s, for most of that time with their two daughters and now a mongrel named Bella—and sat down at the drawing board in his cramped office, its drawers jammed with drawings, its walls haphazardly covered with framed pictures of boats he'd sailed or built, a few shelves loaded with dusty issues of *WoodenBoat* and *Classic Boat.* He had, scattered on his drawing board, numerous "ducks," lead weights with L-shaped pins on them that held down long plastic strips, or splines, and many ships' curves for fairing lines. He taped a piece of vellum to the board, put a mark at one end, and, figuring ⅜ inch per foot, measured 22½ inches toward the opposite side of the paper, then put a mark there. He then struck the load waterline and made a general guess at the draft, how deep the boat would be beneath the waterline—certainly not more than 9 feet for *Rebecca,* and probably not less than 8, he figured—8½ feet.

The width was simple: "Not too wide and not too narrow!" he would explain later. He wanted a beamy boat for stability, comfort, and deck space. "With a boat that has two masts," he reasoned, "you've got a lot of leverage trying to heel it over. So you don't want some skinny, narrow thing. If you do, you have to go way deep with your ballast, and then you've got a quirky, difficult boat." He planned a width of 14 feet or so.

He wanted a powerful bow, and so it would be a little high, and you always want the stern to be lower than the bow—a couple more ticks of his pencil for the bow and stern heights. Above the waterline, on the freeboard, you want enough height so you don't feel like you'll get soaked every time the boat heels a little bit, but not so much that you need a ladder to get in and out of a dinghy.

Now he had the bow height, the stern height, and the lowest part of the freeboard—roughly 4 feet. He set his ducks to hold the batten down at these points and drew a fair curve.

This was how *Rebecca,* a traditional wooden boat, began to take shape in the summer and fall of 1995. On September 20 Nat finished a working sketch, and by the following February he'd completed the lines drawing, a presentation of the hull from four perspectives. The profile drawing tells you what the boat and the rig will look like; the lines drawing tells you what will make the boat tick.

Rebecca was, Nat says, the easiest boat he's ever drawn.

✳

Two observations by people who know Nat come with enough regularity to make them seem his dominant attributes, at least in other people's eyes. The first is that he is "deeply spiritual," or "very spiritual," though no one is ever able to offer any evidence in support; it's just a sense, but enough people feel it to give the observation credibility. The second is that nothing fazes him. He's unflappable. He is almost always happy, and more: he's at peace. Bernie Holzer, a seasoned seaman and a steamship purser who has known Nat for years, says he's never heard him raise his voice.

Nat is a Christian Scientist, and while he's not orthodox in his adherence to the religion, he says he's learned most from its suggestion that we approach problems from a spiritual rather than a physical or material angle. Pam was raised a Christian Scientist, and while Nat resisted learning about it for years, ultimately, the more he read and thought about it, the more Christian Science dovetailed with ideas and knowledge already ingrained in his thoughts. "I think everyone goes on their own spiritual path in their own way and observes or finds what seems to be suitable," he says. "What seems to work for them. I'm a slow learner, but gradually it became clear to me that this was something to reckon with—there is some truth here, and it's worth pursuing."

Nat is so steady that few things seem to impress him overly, and he's not inclined to overpraise, but neither does he seem to take anything for granted. Perhaps he comes by this honestly. At twenty-one he sailed a small wooden boat, a Block Island schooner, across the Atlantic with a single, inexperienced crew member—a hard forty-one-day crossing with a near sinking, a near death, heavy weather. He made landfall in Newport and within hours called his mom—after six weeks

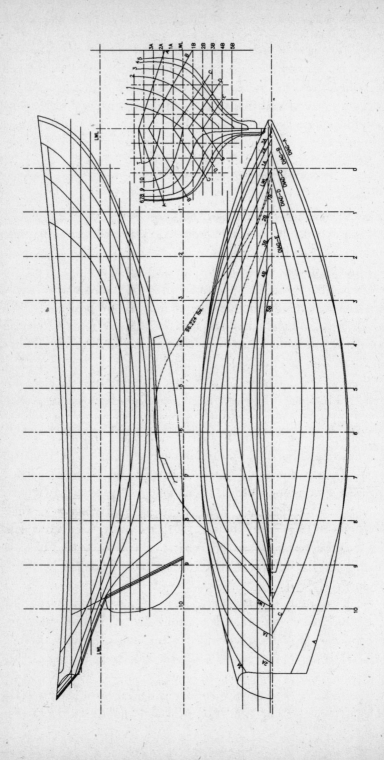

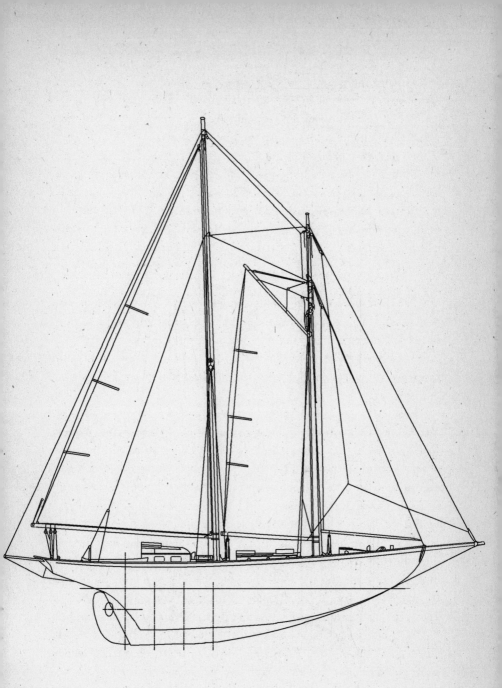

Rebecca lines drawing, *left,* and sail plan, *above,* by Nat Benjamin

at sea, incommunicado the whole time, since there was no radio on the boat—to tell her that he was all right, they'd made it across the Atlantic. Her first words were "Do you have a summer job yet?" Nat was silent on the line for a moment; then he chuckled.

He smiles now and says, "She put me in my place."

So maybe his incredible steadiness, his even keel, was learned, but it is also linked to that first observation people consistently make about him. His steadiness and his spirituality are connected. He is a man who mainly dwells elsewhere—out on the bowsprit or way up the mainmast on a spreader.

✴

When Nat says that this 60-foot boat was easy to draw, you have to remember that he's not easily impressed, particularly by himself. So in order to understand the actual significance of *Rebecca,* you've got to pull back from Nat's house on Grove Street and the earthy island of Martha's Vineyard and look well beyond, until the boat can be seen in her proper context.

Not that long ago, sailboats that were made out of wood were not special because of what they were made of—they were usually made of wood. In 1930 you wouldn't have said, "That's a beautiful wooden boat," as you must do today. That would have sounded idiotic back then, like saying, "This is delicious flour bread." But now, in America and the rest of the developed world, almost no boats are made out of wood. Since the 1960s fiberglass has dominated the field of boat construction. Metal has for more than a century been widely used as well, most notably for big boats (though some sailors do enjoy small steel hulls), but beginning in 1942, the year Ray Greene built the world's first large polyester/fiberglass object (a boat, as it happened), and working up to apparently unstoppable and permanent predominance in the 1960s, fiberglass has been the premier material for building boats. It's cheap and light and can be molded into any imaginable shape—perfect qualities for that complex system of curves and reverse curves that comprise a sailboat's hull. And because the glass is shaped using a mold, it's possible to make a lot of boats quickly. Fiberglass also turned out to be a snap to maintain, relative to wood. A wooden hull you have to haul once a year, blast it clean, then sand and paint it to keep the growth and barnacles off. Old wooden boats you're forever repairing: broken ribs, water raining through the deck onto the bunks, seeping inexorably through the cracks in the planking or through punky spots where the iron fastenings have corroded. Fiberglass doesn't leak. Fiberglass doesn't rot.

If you neglect wood, the wood resents it. Fiberglass couldn't care less. Wood is humanities and the arts, fiberglass is science. Wood is emotion, fiberglass is reason.

And yet a few people kept building boats out of wood in the modern 1960s and even in the 1970s—oddballs, back-to-nature hippies, and eccentrics who just happened to like them. Wooden boats often stick around for a long time, and those tired old wooden boats were cheap for impecunious yachties willing to do a whole lot of work on them, willing to spend more time working than sailing, if they were lucky enough to do any sailing at all (often, floating was as far as they got).

But *everyone* liked the looks of a well-maintained wooden boat—no one who spent time on the water denied that these vessels could be truly beautiful, and there wasn't a boater alive who wasn't openly grateful to those few poor souls, God bless 'em, who couldn't help themselves and owned wooden boats. But that didn't mean you had to own one yourself! No, thanks very much for the offer, but I'll just enjoy the looks of your lovely wooden boat from my no-leak, low-maintenance fiberglass boat! Hee hee hee.

Indeed, an obscure magazine idea, a magazine devoted to wooden boats, became a resounding success precisely because readers didn't have to own wood to love it, admire it, or even dream about it.

No firm numbers are available, but industry experts guess that fewer than 10,000 wooden boats exist in America, not including dinghies, canoes, kayaks, homemade plywood skiffs, and the like; more likely they number in the mid-four-figure range (6,307 boats are listed in the 1999–2000 edition of the *Register of Wooden Yachts*). There are fewer than 750 boatyards in the country that in some way work with wooden boats. Yet this minuscule industry—several thousand boat owners, roughly 750 shops—generates a subscription base for *Wooden-Boat* of more than 100,000, making it one of the biggest boating magazines out there.

So clearly, plenty of people loved and cared about wooden boats all along, even if they didn't own them, and for others wooden boats were the only kind they could afford and maintain. For these reasons, a few people continued to build them, though in almost immeasurably small numbers, and a few apprentices learned the trade from them even during the wooden boat nadir of the mid-1970s, when fiberglass and plastics were the rage.

But as the 1970s gave way to the 1980s, so many things were being made of plastic—it had become so pervasive in industrialized culture—that "plastic" turned into a metaphor for cheapness, impermanence, being fake, phony, imitation, not the real thing. Wood was the opposite. People returned to wood in their furniture, in their children's toys; they yanked up that ridiculous shag carpeting and sanded and varnished the tongue-and-groove oak boards beneath. They returned to a lot of natural things, as could be witnessed in the trend toward organic farming and eating unprocessed foods, for example, or the preference for wearing cotton clothing after the fads of synthetic fabrics. And they bought wooden boats. Contemporary culture was getting a little too fast and—in this great capitalist economy that heaped riches on whoever could sell the same item for the least amount of money—a little too cheap. Furthermore, fiberglass, it turned out, wasn't a miracle material after all. You *did* have to take care of it. It blistered and cracked; it didn't much like salt water or sun. And if you hit a rock hard enough, or heavy seas pushed you into the shore, that light, inexpensive, easy-to-maintain fiberglass hull could shatter like an eggshell. At sea it bent and buckled and fell apart.

A sailor named Pete Goss recalls such an experience in a 34-foot fiberglass sloop, in a storm in the middle of an Atlantic crossing, in his book *Close to the Wind:*

> *Poor old* Sarie Marais *gave her all for the twenty-four hours that the storm blew. However, it was just too much for her and I suspect we pushed her a bit too hard. First, the structural frames in the bow cracked—to such an extent that the bow section flexed so much that the forehatch kept springing open, flooding the boat with gallons of water. Next, cracks appeared in the deck—first by the chainplates, and eventually running aft along the deck for about six feet. By now the hull was*

flexing so much that gaps of up to an inch were opening and closing near the bulkheads and we had to be careful where we put our fingers. A large split developed in the hull under the engine and the rudder felt loose. It was a hard storm. . . .

As the trip progressed, Chris and I continued to bail, gradually clearing the boat. Then, to our horror, we discovered the source of the leak now seemed to be in the area around the keel fastenings rather than the forehatch. We feared the damn thing was about to fall off. . . .

The boat was flexing so much by this time that if you sat in the companionway with your elbows resting on either side of the hatch, they would go up and down by about an inch with each flex. Our diet was dictated by whichever cupboard or drawer would open—most were jammed shut by the distortion of the hull.

Goss and his companion finally made landfall in Newport, and when they lifted the boat out of the water to check the damage, the keel fell off.

This boat, or any other like her, didn't get fixed. You put a hole in a fiberglass boat, you didn't patch it, you didn't try to save the boat, you just threw her away and got a new one: such boats were disposable. And you never tried to *save* an old fiberglass boat—what was the point? Broken or old, she was only plastic—dump her. It wasn't like she was a great old wooden boat, a boat with *soul.*

In the boating world, wooden boats, or at least those wooden boats that were properly cared for, became a kind of symbol of the natural world, of those things that were good and true, an antidote to the crass, cheap, commercial culture all around us and on sale in strip malls and Wal-Marts.

In the 1990s wooden boats likely began to be made in somewhat greater numbers than in the previous two decades, because of the trend toward the natural and because of a new process that, though arguably more than a century old, had only now hit a critical mass in the wooden boat world. Wooden boat builders, thanks to the fiberglass and chemical industries, took advantage of new, highly sophisticated adhesives and applied them to wood in a method called cold-molding. Boatwrights found that by building up thin layers of wood and glue, they could make hulls that were stronger than fiberglass but lighter than traditional wooden boats, which tended to be heavy and slow. The process didn't require huge timbers (which were in increasingly short supply in this country) or heavy machinery. In short, this method of building laminated hulls combined the best qualities of fiberglass (workability, lightness of material, and no leaks) with all the advantages of traditional plank-on-frame construction (durability, beauty, and the sensory pleasures of sailing a wooden vessel—her sound, smell, and feel as she charged through the waves, the solid lap against the hull at anchor, you in your bunk on your way to sleep).

Within the wooden boat world, there was hot debate over whether these new cold-molded boats should be considered wooden at all (the traditionalists dismissed cold-molded hulls as "wood-reinforced plastic"). But the facts were these: one kind of boat was made of wood held together by metal screws and bolts, and the other of wood held together by glue. Both were, in fact, wood. This position was wholly embraced by the arbiter of this world, *WoodenBoat* magazine.

And so with traditional builders still working, and many boatyards picking up new commissions for cold-molded boats, the wooden boat industry appeared to be in fine health, chronicled bimonthly in the gorgeous, glossy pages of *WoodenBoat*. There were nonprofit projects here and there for which preservation societies built large wooden boats, often replicas of sailing ships of bygone eras—like *Pride of Baltimore*, for example, or *Amistad*, a replica of the slave ship recently launched at Mystic Seaport in Connecticut. But over and above this general robust health, three significant projects appeared to give the sense that something more was happening than a little blip of increased activity in the wooden boat world.

✸

In 1997 sixth-generation builder Harold Burnham launched in Essex, Massachusetts, *Thomas E. Lannon,* a 65-foot schooner commissioned by Tom Ellis for pleasure sailing and chartering. This heavy plank-on-frame vessel, modeled on Essex fishing schooners of the late 1800s, was among the biggest traditional vessels privately built for fun and profit since the advent of fiberglass.

Next, Donald Tofias commissioned two 76-foot, cold-molded racing boats designed by Joel White, one of the preeminent designers in the country (and the son of the celebrated writer who created one of the great fictional yachters in the canon of American lit: Stuart Little, the mouse). These boats were to be built consecutively by White's son, Steve, at the Brooklin Boatyard, and his son-in-law, Taylor Allen, at Rockport Marine, both on the rocky coast of Maine.

Tofias's plan was not to *own* two sleek wooden racing yachts, but rather to sell them and create a class of huge racing boats that people all over the world would buy and then race in various locales. This latter practice, called one-design racing because it doesn't require the complicated and often controversial handicapping of boats, was all but dead by the time the Great Depression struck, and Tofias argued that it in fact hadn't been done with any large boats since Nat Herreshoff designed and built the New York 50s in 1913. Whether or not this appealing, perhaps hopelessly romantic, and certainly expensive venture (each boat would cost a million to build and sell for twice that) would pan out could not be known for years after the boats' launchings, in the summer and fall of 1998. More boats and plans were in the works; Tofias didn't think small. His goal was to create five classes of boats, ranging from 46 feet to 130.

Whatever the ultimate fate of Tofias's W-Class yacht company, the first two 76-footers would be built, and the remarkable fact of the matter was this: for the first time in nearly a century, a man was *investing* in wooden racing boats, considering them a commercial, for-profit venture. "We want to build a lot of racing boats over the next ten years!" Tofias proclaimed to the applauding crowd at the launching of the second boat.

And finally, *Rebecca. Rebecca* combined the most exciting aspects of each of these projects and, in the minds of many, surpassed in ambition and scope any other current construction, or any construction anyone could think of or remember in decades. *Thomas E. Lannon* was a big, fine schooner, and traditional, but she remained true to her workboat roots; she did not have the exciting design of *Rebecca,* a deep-ocean yacht, a big schooner rig with a powerful bow that would barrel toward you, looking every bit the mighty boat that she was, and then slip past and disappear with her fine, graceful stern. And she wasn't cold-molded. Many still debated the veracity of the claim that cold-molded boats were wooden boats (at the launching of the second

W-class, *White Wings,* in Rockport Harbor, Ross Gannon nodded in
her direction and said to me, "That's not a wooden boat"). *Rebecca* was
pure and true. *Rebecca* was a great traditional wooden boat, with traditional lines, the kind of boat that hadn't been built since before World
War II and the dawn of fiberglass and the cultural transformation that
had opened up yachting to the masses of the middle class (with fiberglass, almost anybody could own a boat). *Rebecca* was simultaneously a
link to the golden past of sailboat construction and a daring, hopeful
signal of the future.

II

About a year after Nat Benjamin completed the first lines drawing of *Rebecca,* a boat named *Sanderling* was floated at high tide alongside the dock off the Gannon & Benjamin beach. A rusted Hathaway winch, a big old monster discarded from a Connecticut boatyard that G&B had picked up secondhand and bolted into the tracks, slowly released a platform cradle, made of timbers attached to I-beams on wheels along rails, rails that descended down the beach and disappeared into the water as if built for an Atlantis-bound train. *Sanderling,* a 33-foot Malabar Sr. built in 1957, was maneuvered over the half-submerged cradle; four large wooden arms, 'thwartship timbers, were hoisted up against her hull and held in place by steel pins. The winch then hauled the cradle back up out of the water, onto the beach, and the boat came with it. The winch looked like something from another era—and it was, but it remained a powerful piece of machinery, its inch-thick cable capable of hauling, on this beach, a boat weighing 50 tons. *Sanderling* needed major work: the forward section of the keel, called the forekeel, needed to be replaced, as did engine bed bearers, many frames, the sheer clamps (long planks that run the inside length of the sheer, important structural pieces of a wooden boat), the deck beams that rested on the sheer clamps, and the deck that rested on those beams; a rehab of the accommodations was scheduled; and last, a new engine, along with mechanical and electrical systems, had to be installed. It would be a full season's work, requiring several boatwrights.

Sanderling would sit for many months on blocking in the sand be-

side the shop itself. The G&B boat shop, a long, rectangular structure of weathered cedar siding—genuine New England rustic—was
divided into two main parts, of about equal size. To drive to Gannon
& Benjamin Marine Railway, you turned north into the driveway,
passed the offices of the *Martha's Vineyard Times,* and stopped in back
by a green Dumpster, or nosed your bumper up to the friction winch
in this parking area with space for three cars, which typically contained
ten or so at any given hour of the workday. You walked up the steps
and through the sliding white barn door, on which hung a flower box
decorated seasonally by Ross's mom. If the day was sunny, you would
be, upon entering, disoriented by darkness. You might bump right into
the table saw, a fairly nice piece of equipment, maybe the yard's finest,
and about the only freestanding piece of machinery in the place that
had been built after the Hoover administration. To your left would be
the planer and jointer, which had the comforting look of a 1920s
cast-iron industrial piece of work. The jointer no longer functioned,
but the planer was operational; the whole building trembled when a
plank was sent through it (protective maroon earphones hung on the
thickness-adjustment crank). Behind this sat an enormous ship's saw, a
great big band saw whose blade tilted up to 45 degrees while running.
Beyond that, against the west wall, was the Hendey metal lathe, circa
1930s, cast iron like the planer, originally built for the Brooklyn Navy
Yard and formerly steam-powered.

Scarred workbenches stood beyond this, apparently fossilized,
and above them, against the far wall, various drawers on which were
scrawled barely legible words such as *Tapered drill bits, Planer blades, Plug
cutters,* and *Ross Engineering,* for Gannon's special handmade tools. Much
of the eastern wall, past two smallish, haphazardly placed band saws,
was given over to similar drawers, though if you were looking for the
number 14 bronze screws, they'd just as likely be spilling out of a box
or a brown paper bag on the workbench below, which ran the length
of this wall. The wall had windows that were always black; beyond
them, on the outside of the building, were racks of gaffs, booms, and
masts, covered for the winter by recut old sails against the rain and
snow whipping off the Vineyard Sound. On the wall, covered with
sawdust, were snapshots of employees and friends sailing in sunny
climes, ripsaws, an old radio, chisels, hand planes. Most of the vises on

the workbenches worked. Close overhead was nothing but odd lengths of wood battens shoved up on racks, coiled band-saw blades, electrical wire.

And that was it—before you knew it you'd have bumped back into that nice new table saw and would be puzzling over why it looked so out of place. You'd only gradually become conscious of the fact that everything *but* the table saw looked ancient. Machinery dark with grease, and every surface not just coated with sawdust but caked with it—these gave every object, most of which were pretty old to begin with, the appearance of decades of disuse. Indeed, the sawdust was so plentiful that it made the whole shop look like the interior of a sunken ship viewed in grainy black-and-white documentary footage, the artifacts of a past era thickly encrusted with barnacles and coral.

Countless strangers and tourists, fresh off the ferry five hundred feet down the road, toured the shop just this way, as if the boatwrights and the place itself were an exhibit in a museum.

While the summer hordes passed through anonymously and with little effect, the people who lived here year-round revered and admired Nat, Ross, and their shop. Gannon and Benjamin had created a productive waterfront business within the local economy, but more important, they were a spiritual part of this island, which prided itself on its rustic, nautical heritage and on rugged self-reliance. Martha's Vineyard—in spite of the surges of media that would descend on it sporadically, the increased house construction for the very rich, and escalating tourism—was beautiful, and huge stretches of it remained all but untouched thanks to fierce conservation measures and building caps (though many thought these efforts were too little, too late).

Vineyard Haven was officially the port of the town of Tisbury, its one main street featuring shops and small restaurants and a movie theater, a great independent bookstore called the Bunch of Grapes, a drugstore, and a bank—enough nonboutique businesses to keep it from being quaint. Just north of Main Street was the ferry landing, the single year-round place of arrival or departure by sea. About three hundred feet down the beach, next to Bob Douglas's Black Dog Restaurant and the Coastwise Packet Company, was the G&B boatshop and dock. They sat at the very crook of the large V whose extremities were formed by the strips of land called East and West Chop,

Vineyard Haven Harbor being the last small portion of this V, separated by a breakwater from the outer harbor. The harbor had remained, against odds, a genuine working one, its westernmost portion spotted with boats on moorings—mostly wooden boats these days, thanks in large measure to Gannon & Benjamin, which had set up shop in 1980 and was still sometimes referred to as "that hippie boatyard," or "that funky boatyard," though Nat and Ross themselves were well into conservative middle age. Deeply entrenched in the community, a solid and respected business, it had retained its ramshackle appearance and the pleasantly haphazard, cluttered feeling of an old country antique shop. Most of its contents *were* antiques, even if they were in use most days.

Ginny Jones managed the affairs of the yard. Her office was just off this first section of the boatshop. Her door had a wooden latch and a port light in the top center. Ross had built Ginny's little four-by-eight-foot space to resemble a boat's chart station, and when you were in it, you could imagine you were on a boat. Two people could work in here not uncomfortably; three people was acceptable if all were engaged in the same matter; sometimes four people and a dog would be in the office, but not for long. Its back wall was all mullion and window-pane, the only reason the cabin didn't feel claustrophobic.

Ginny called her ancient brown Peugeot, its spare tire lashed to the rooftop, the Battle Wagon; some might reasonably call her the same. In her midfifties, with a long, dark rope of hair secured in a ponytail, glasses, a round face, Ginny carried the weight that bespoke her talents as a cook (finding herself in possession of a side of Alaskan salmon, she'd poach the whole thing, bring it in, and set it on the wood-burning stove between the planer and the ship's saw, and people would not only devour the thing, they would talk about it for as long as a year after). Ginny also happened to be boatstruck and would have worked almost anywhere that allowed her such a view as she had: from her seat before a slim wooden work surface just deep enough for "Pandora"—a laptop computer with which she, a Luddite, maintained an armistice that she described as "armed neutrality"—she could gaze out at three of the country's most extraordinary wooden boats floating in the harbor, *When and If, Shenandoah,* and *Alabama,* the latter two 108 and 90 feet. She would often conclude an e-mail to a colleague by describing their position: "It's a beautiful afternoon and the three big schooners are all facing NNW. They've got a rim of ice along their waterlines . . . ," or "A beautiful full moon high in the sky over *When and If . . .*" For her it was the same as saying "All is well" or "Yours truly." And then her signature: all missives were signed "The Madam." Ginny Jones, Virginia Crowell Jones, was the bookkeeper, the finder of obscure parts, the manager, head lobbyist, Steamship Authority gadfly, commentator on boatyard gossip, bill payer, paycheck writer, chief grumbler, lover of wooden boats, and, by her own description, resident pinko. A true Vineyard-born girl.

Through two doors, to the left of Ginny's office, you entered the other half of the boatyard, which had only three walls, with the eastern edge of the floorboards dropping off to sand and a view of the docks along the beach and the far side of the harbor, where Ralph Packer's huge white oil tanks sat like industrial monoliths beside the Martha's Vineyard Shipyard. In this open section of the boatshop, new boats up to 32 feet long were built; the floor was painted white about once a year for lofting, the drawing of a picture of the boat's exact measurements and contours—in effect, a full-size blueprint of the boat that would materialize piece by piece above. Currently the shell of a 21-foot sailboat named *Blue Rhythm,* a sporty little gaff sloop, grew here; Nat's design, called the Bella class, would prove so popular that three more would be built almost immediately after the first was launched. A workbench ran the length of the long wall of this section of the shop, and an old yellow refrigerator pocked with rust stood at the far end, near a paintbox and other painting materials; at the near end sat a big metal welder and a fourth band saw. Between these items was endless clutter, below the workbench and above, hanging from the walls and roof (another welder dangled from a rafter above the band saw, tied up like a Christmas package alongside an old rowboat, paddles, various boat lines, unwanted oil paintings, and bags stuffed with sails).

Stairs ran up the back wall to what was called the Loft, over the enclosed half of the shop—a sail loft run by Gretchen Snyder, who made sails, seat coverings, and anything else of fabric that a boat would need.

Gretchen, who had spent nearly half of her forty-five years in the Loft, came and went all day long, ever happy, smiling and laughing a

winsome laugh so melodic it sounded almost like birdsong. She invariably wore a short dress, cotton stockings, and, for most of the day, knee
pads. Gretchen was a sailor, too: she taught sailing in Maine in the
summer and took two vacations every year to deliver a boat called
White Hawk from New England to Antigua in the fall and from Antigua to New England in the spring. On her way in and out of the
shop she would pass *Blue Rhythm* and there, on the blocks beside the
boatshop, was *Sanderling,* whose cruising strength and wooden beauty
day by day through the seasons were slowly restored.

✳

The man in charge of replacing *Sanderling*'s forekeel was Brad Ives.
Brad was a new employee but an old friend, having met Ross in the
early eighties in the Caribbean, where Ross had chartered boats in
the winter months and Brad had delivered tropical hardwoods on *Edna,*
his 100-foot steel-hulled ketch. Shortly thereafter, G&B had begun to
build boats using such South American species as silverballi, wana, and
angelique, the last a species so hard and dense that rain-forest bugs can't
penetrate more than an inch into its dark-brown trunk, which is an
inch farther than you can drive a nail into it. Angelique is like cement,
extraordinary material to use for keel timbers, garboards, and sheer-
strakes, parts of a boat that need to be very strong. Silverballi and wana
are softer woods and make excellent planking stock. The best forms of
silverballi dry quickly and are exceptionally stable, and golden in color;
wana, similar to mahogany in appearance, deters marine borers and
makes excellent planking stock.

Brad soon left to sail in the Pacific, and the boatyard didn't see him
for several years. He returned for a half year in 1994 to work on the
restoration of *When and If;* then in 1997, when his work was concluded on a boat called *Picton Castle,* a 150-foot bark in Nova Scotia,
he contacted G&B. He'd been in touch with Nat about the schooner
project, and G&B told him to head down to the Vineyard.

Brad was a tall, slender man, about six foot two, with fine, straight
black hair, high cheekbones, a small, sharp nose, gaunt cheeks, and a
prominent chin—a narrow face, full of edges and hollows. He moved
with an almost exaggerated slowness and spoke quietly and infre-

quently, even among the other boatwrights. He did his work. Wasn't prone to smile. When visitors appeared, especially those with pen and notepad—reporters from the *Martha's Vineyard Times,* the *Vineyard Gazette, WoodenBoat,* and the *Boston Globe* frequently materialized to interview Nat—he would recede and avoid inquisitive strangers, occupying himself with his solitary work. The first time he saw me, I happened to have my notebook open, and he said, "Jeez, another *writer?*" Brad from the outset was mysterious to me, like the hired hand in a Western, the one with A Past. I asked Nat about him, and he suggested that Brad might not be forthcoming, hinting that he didn't care much for writers and that he'd had some trouble with the law a while back that maybe he didn't want to talk about. But Brad never failed to answer my questions, and he said he had nothing to hide—to him the main disadvantage of strangers, apparently, was that they took you away from your work.

Brad was unusual at the boatyard for being the only one who didn't do much woodwork. He quickly became the mechanical-systems guy because, though more than capable in all areas of boat construction, he didn't have a natural touch with wood. This became evident when he tried to put in the new forekeel, a big piece of the boat, in *Sanderling:* it didn't fit, and he had to cut another. Brad liked big sailboats—90, 100 feet—and with a boat that big, he wanted a steel hull for strength. He had once nearly lost a boat, a 90-foot Baltic trader, to a hogged keel that popped out the garboards (the planks fitted into the keel), sinking her in shallow water. Repairs on the shores of Costa Rica had taken nine months. "I like steel boats," he said. "I like bigger boats. Wood gets to be a problem with bigger boats. You can do it, but the strength's not there." You could fix a hole in a steel hull easily with a blowtorch and scrap metal—he liked that. Moreover, he saw beauty in steel hulls, in a line of rivets as fair as a sheer. He was the metal man in the wooden boatyard.

You'd never know, glancing at Brad, that he'd spent most of his adult life delivering cargo under sail in big boats and sometimes encountering more adventure than even he had wished for. You'd likely see him below decks on a hauled boat, focusing through reading glasses at a tangle of wire in his hands. He came and went as he pleased, did his work. Most in the yard knew parts of his history, though only in a vague way, a story here, an anecdote there. But when in the actual presence of the quiet, even, unsmiling Brad Ives, they all felt an unmistakable depth and force of intelligence, a wisdom born of experience at sea.

Ted Okie, an apprentice in his midtwenties—with wavy blond hair, a ready smile, a scrappy beard—who would later come to work at the yard, would say of Ives, "If we were more primitive, we'd call him a magician or a sorcerer."

✳

That winter, early in 1997, as Brad was installing the forekeel, twice, on *Sanderling,* Nat had a talk with him. Nat had contacted Brad as soon as he began sketching *Rebecca,* and now he told him that it looked like the schooner project was going to go forward. Brad understood the magnitude of the project, and it was a main reason he'd called G&B about work. Negotiations with the future owner had been progressing hopefully. The reason Nat had been in contact with Brad was, of course, his chief concern. This was a big boat. There would be thirty, maybe forty thousand pounds of wood in *Rebecca*—not raw lumber but finished, hand-cut pieces. They'd need double that weight, twenty or thirty *tons* of wood at least, before they pulled the start cord on the chain saw to cut out the first keel piece. That was a hell of a lot of wood. Cost was a factor. Availability of wood in long lengths was a factor—the longer the planks, the stronger the hull. But the biggest issues in Nat's mind were the wood's quality and longevity.

Rebecca was the boat of a lifetime, a projection of so much that Nat loved in sailboats—size, rig, traditional lines. This could be his crowning achievement as a boatbuilder. It was monumental for him, for the boatyard, and for wooden boats generally. He might never have an opportunity like this again. He would not use mediocre wood for this boat; he wouldn't even use very good wood. He wanted the best available on earth, and a lot of it. A year or more of many people's lives would be devoted to building this boat, and it would be criminal to use material of a lesser quality, to have them hand-carve it piece by piece: it would be a waste of their lives and their talents as craftsmen. Furthermore, it might literally be criminal if, once the boat was built and found herself in heavy weather, she fell apart because the wood strength was insufficient—whether a year after launching or fifty or a hundred years after, for Nat knew that a great wooden vessel could last that long if she was cared for. And he knew also that if her material was the finest in the world, it was more likely that she *would* be well cared for. Nat was, perhaps more than a builder, a sailor, and he was thus gravely aware of the stresses the sea put on a boat that size, a rig that large. People's lives would depend on her strength and soundness more than on her lovely brightwork and fancy finish carpentry. People died out there because of bad materials, cheap hardware, faulty construction, a lack of care. Great wood in wooden boats ensured its own life by inspiring great work and reverential care, and it protected the lives of those who would sail those boats because good wood meant durable wood.

But wood of this sort wasn't easily available in the United States. Not in the size or at the price or of the quality he'd need. Remarkably, though, the one person on earth he knew who might conceivably be able to find such wood happened to be working for G&B. "For the planking and the stock that he gets," Nat says, "I don't know anyone else who could supply it."

And so he and Brad resumed a conversation that had been going on for a year already, ever since the lines drawing of the schooner was finished. Brad explained that he hadn't been to Suriname—a country that he knew had the kind of wood Nat needed, and had it in abundance—since the mid-1980s. Civil war and military rule had ended, Brad said, but what shape the country was in, he had no idea. Formerly Dutch Guiana, and situated between Guyana and French Guiana on the northern coast of South America, Suriname was scarcely known in the United States, and he'd had little news from there since he left. The lumber concerns he'd been working with, were they even in business anymore? He didn't know. Was the economy in shambles? Probably. Was the government corrupt? Probably. But he was willing to fly down there and find out what he could get, if someone could front him the money for the wood and pay him a hundred dollars a day in wages and expenses. If that could be worked out, he thought he could get the wood Nat wanted. He and Nat met with Dan Adams, and Dan agreed to bankroll the trip.

✴

Brad took the four flights required to reach the airport outside Paramaribo, the capital of Suriname, to begin several months' work in this Third World country. It had indeed changed since he'd last been there—storefronts once filled with new European goods were now empty and dark, roads were crumbling, some of them with decade-old bomb holes, and city sidewalks went unrepaired. Nevertheless, he was excited to be hunting wood for G&B and a big wooden boat. Jump-starting his importing business, furthermore, enhanced his sense of a new beginning, a beginning that would have been impossible during his three and a half years of parole, which forbade his leaving the States—he who had lived most of his life separate from any govern-ment, needing to honor and respect only the laws of nature. He was free. It was the beginning of a new phase of his life: his daughter, Wil-low, was grown and was soon to head off to college; he'd begun a new, happy relationship with a woman named April; new, steady work at G&B on a monumental wooden schooner completed this fine picture and hopeful life.

And despite the upheavals of a decade's unrest, he did rejuvenate good contacts here. This was critical. Nat had ordered thousands of board feet of silverballi, in long, wide lengths, every inch of planking for a big wooden ship, and massive timbers for its backbone. The Suri-name rain forest, one of the most untouched in the world, with one of the most diverse ecosystems, by and large remained that way, except for vast chunks that huge concerns from Malaysia had been allowed to clear-cut. The government was just becoming aware of the value of its forest, and this was a main concern of Brad's—he would use only re-sponsibly cut trees, and so few of them relative to the entire forest that government officials monitoring his exports would judge his take to be negligible.

But no one was cutting the wood he needed, and so he had to coax the yard owners and cutters—the Bush Negroes (descendants of Africans enslaved by the Dutch who had immediately fled to the forest, still thriving today) and the indigenous Amerindians—to locate excel-lent silverballi trees in the dense rain forest, cut them down with chain saws and drag them out with trucks called skidders, and then float them down the Cottica River to the mills.

He kept an eye out for pieces of wood that were curved. A strong curved piece of wood is useless to a carpenter, a cabinetmaker, a house builder, but it's treasure to a boatwright. A boat is composed mainly of curves, and if a piece of wood has grown with a curve in it, that curve will be stronger than one manipulated by bending or sawing.

Early in his stay Brad spotted a log of angelique at the Oemrawsingh Sawmill measuring 23 feet in length and 28 inches in diameter. It had been at the mill for a while because nobody wanted it—it was one long, gentle curve, and nobody liked sweepy logs. Brad scrutinized the unwanted piece of wood, carefully examining the cut ends for signs of heart crack. He measured the log and again regarded its graceful sweep. You never knew what you were going to find on the inside of a log, but if this one remained as good as it looked at either end, it might just work for one of the most important pieces of *Rebecca,* the forekeel—the piece of wood, much on his mind recently at the boatyard, that connects the stem to the main keel timber.

Brad directed this curved log to the horizontal band saw, where two sides were squared off. He then took a couple of 1½-inch planks off the top. The interior looked good. They rolled the log and took a couple more off the other side. The interior continued to show straight, regular grain with no cracking. Brad now had a piece of wood a foot and a half thick and nearly two feet wide, with a twenty-three-foot sweep, weighing approximately 3,500 pounds. A single piece of wood. The log was nothing special as far as logs went, except that it fitted exactly the designed curve of *Rebecca*'s forekeel—it would be nothing special to most anyone who used wood, that is, but it was as close to a gift from God as a boatbuilder could wish for.

Rebecca might be defined simply: pieces of wood held together by bronze fastenings that would stay together under dynamic conditions at sea. Those pieces of wood, each one, would be hand-cut and fitted into place, whether a plank, a frame, or a deck beam. And few in the boatyard would lose sight of the fact that this hand-crafting had begun at a mill in Suriname, or that the hands in question had been those of Brad Ives.

Brad spent five months locating the trees and milling the wood—hundreds of pieces amounting to a hundred tons of dense, heavy lumber. What he didn't know was that there was no longer any money to pay for this wood.

To finance the construction of *Rebecca,* Dan Adams had set up a company called Mugwump Charters, Inc., and given Gannon & Benjamin access to a business checking account. The deal was this: Dan would deposit money into the account, and Gannon & Benjamin would use it to pay for materials. This way, when Brad was ready to purchase the wood, Ginny could send him American cash immediately. When that time came, however, and Ginny went to wire Brad money for the wood, the account was empty. *No cash. Gone.* And there was Brad sitting with many thousands of board feet of the most amazing boat timber available, cut to lengths and widths that were scarcely available in the United States.

Dan had neglected to tell Ginny or Nat or Ross that he, too, had a means of withdrawing funds and had needed the money, all of it, for something else. He had simply neglected to mention this fact. Ross confronted Dan in the boatshop. *You can't do this,* he said. Dan, casual as could be, said, *It's going to have to happen from time to time.* And Ross replied, *No, it cannot happen. You can't build a boat that way.* And Ross was right.

Brad got paid, of course—Dan came up with the cash, or most of it—but it was ominous behavior on the part of the man who could end the construction of *Rebecca* if he so chose. The little glitch was soon forgotten, though, and Brad never even knew there had been a problem in the first place. He simply did his work—finding the wood, cutting it, then ensuring it was loaded properly into containers and shipped out of the mouth of the Suriname River to Fort Lauderdale, where he met it and got it through inspections. He then saw to it that the wood was properly loaded onto the eighteen-wheelers that would deliver it to Martha's Vineyard.

Brad was still loading the last trucks when the first of them called G&B from Woods Hole, asking for the ferry fees. Ginny Jones would run a check over to the Steamship Authority, which ran the ferries, and the eighteen-wheeler would be floated across to Vineyard Haven, then would rumble down Beach Road, past the ramshackle boatyard, to un-

load the wood on Ralph Packer's lot. Brad had left in May for uncertain territory, but by the following November it was clear that he'd
come through, and done so royally, sending up five trucks, each carrying twenty tons of some of the most extraordinary boatbuilding timber
on earth.

For Nat Benjamin, it was the most thrilling moment of the project
so far, seeing these trucks arrive. He grinned, his teeth clamped. *He*
didn't see a hundred tons of wood, of course. He saw *boats.* Not just
one boat, a great big schooner, but a slew of plank-on-frame wonders.
Ross was building his own boat in the woods back at home, a husky
44-foot sloop; Duane Case, one of the G&B boatwrights, had designed a 38-foot motorsailer for himself and his wife, Myrtle, to live
and cruise on; in a year's time G&B would begin its first powerboat, a
32-foot yacht of Nat's design, inspired by workboats of the 1930s; and
at least two more Bellas were scheduled to be built. In a world that had
seen plank-on-frame construction diminish nearly out of view, this
single delivery of Suriname hardwood felt like a new beginning.

III

T o build a boat as big as *Rebecca,* the builders needed a place to build it *in*. Nat and Ross did not have such a thing. The space they rented, their little spot of beach, was usually squeezed to capacity with boats under repair and under construction. They'd fixed *When and If*—the 63-foot schooner built for General Patton and smashed on rocks in Manchester Harbor, Massachusetts, during a bad blow—beside the boatshop, but they couldn't tie up such valuable space for an entire construction, estimated to take two years. So Nat and Ross found an open area behind the Tisbury Marketplace, down the road a few steps from G&B, a faux–New England strip mall of shops and service stores. They quickly made arrangements to lease the lot and ran a proposal through the zoning and conservation boards.

Nat and Ross met with Dan Adams to discuss the cost of building a gigantic shed. The cheapest thing they could raise would run twenty grand. Dan agreed to it. The main timbers would be taken from his own newly purchased land, which he wanted cleared anyway, and the rest of the material could be ordered from Home Depot, using a credit card.

Ross returned to his house-building efficiency and within seven working days had raised a 2,100-square-foot clear-span pole barn, seventy by thirty feet, ten feet longer than the intended boat and more than twice as wide. The walls were fourteen feet high, and a corrugated metal roof extended upward from there, with open windows running the length of the eaves to let in plenty of light. A door big enough to slip a finished 60-foot schooner through was made in front, and an old canvas sail was rigged like a giant window blind to enclose the doorway. Brad wired the building for electricity.

Ross and Nat had been skeptical from the outset, and the monkeying with the checking account while Brad was in Suriname had enhanced the sense that something here was not right. This *Rebecca* project was no more substantial than the rumors surrounding it. But when the big pole shed was built, Ross could say confidently, "This job is going to go forward." What reason was there to believe otherwise? There was the big pile of wood over there, and there was a big shed right there!

The crew broke for Christmas and then, in the first week of January 1998, put the lofting floor down, plywood painted white to draw a boat on. Workbenches were built, as was a small office in back, and tools were gathered.

Ross collects heavy antique machinery the way other people collect Fabergé eggs or snow domes, and he happened to have an industrial cast-iron planer, originally steam-powered, in his yard under a tarp ("It's worthless to most people, but for us it's just perfect," he says). They hauled that over to Mugwump, the official name of the building and the name Dan Adams had given to the corporation he'd created to build *Rebecca*. They picked up some big band saws cheap because they didn't work, and Brad repaired them, scavenging what parts he needed. David Stimson, having closed his own boatshop, had arrived that fall from Boothbay, Maine, with his wife, their two sons (ages eleven and thirteen), a table saw, a small band saw, and a finish planer. The Stimsons made their home in an old corn shed behind Dan Adams's barn; David would be one of the key builders of the project. Mugwump bought a minimum of hand tools—a Skilsaw, an electric planer, drills, and auger bits. Nat and David began to loft the boat. By January 15 they were taking a chain saw to the main keel timber and then to that curved piece of angelique, ripping out the first pieces of the boat. *Rebecca* was actually under way.

✴

Of all the boatwrights at G&B, David Stimson may have been the one most representative of the contemporary wooden boat builder, the quintessential solitary worker constructing relatively small, traditional

boats way down a bumpy rural road in Maine, a man who did the work
without regard to profit because it was what he loved. Maine was simply where you did it. No other state was more closely associated with
wooden boats. Wooden boats, whether yachts or workboats, were not
unusual there, as they were most everywhere else. That ensured at least
a modicum of business in repair and, with luck, new construction
every now and then. *The Directory of Wooden Boat Builders and Designers*
listed eighty-eight separate concerns in Maine; the state with the next-
highest number in the *Directory* lay at the opposite northern corner of
the country: Washington, with forty-seven yards listed.

David was of medium height and build, with dark, wavy hair, a
full, fine-whiskered beard, glasses, and a clarity of complexion and
gentleness of demeanor that suggested he was a decade younger than
his forty-three years. He loved physical work; the hours that needed to
be spent sanding a boat hull each spring, an obligation most found
onerous, were to him pleasantly meditative.

The work was precarious at Stimson Marine, Inc., where new construction was limited to boats under 22 feet, generally skiffs and kayaks.
But he felt glad for the work he had—he loved those boats. "I was
lucky," he told me. "Others do house carpentry." He paused. "I just
starved." Winters were particularly difficult, especially given that the
nearest school was thirty-seven miles away, and his wife, Tamara, had to
drive the boys there and back daily; over the rural roads in winter, the
trip exceeded an hour each way. And both of them knew that David
was just scraping by—neither could predict from one year to the next if
he'd be able to continue. David described the boatyard dynamic this
way: "There's always a job that had to be finished yesterday and not
enough money to pay the bills." This resulted in more stress than boats.

David, born and raised on Cape Cod, had known of G&B almost
from its beginnings, and he and Tamara had remained good friends
with Nat and Pam through the years they were in Maine. In the summer of 1997 the couple had visited the Vineyard for some sailing and
to see friends, among them Nat and Pam. Nat showed David the drawings of *Rebecca* and said, "It looks like we're going to start work on this
in the fall." Without thinking, David replied, "Wow, do you need any
help?" Nat said that indeed he did—would David be interested in
helping to oversee the project?

By the time David and Tamara crossed the Maine border on their drive home, they'd decided to do it: close down the shop (despite the fact that they'd just bought a 30-foot sloop he was fixing, and he had a big repair job that had to be finished), find homes for all their animals, make arrangements for school for the kids, pack up their belongings, and move to an island, all in a month and a half. But the decision wasn't difficult: it was an opportunity for David to learn from the combined experience of Nat and Ross and to work on a schooner of that size, and more, the long-term prospects were promising.

"They're on a roll here," David reasoned. "There's money here. It's a perfect place. There's a level of culture here that you don't find in other places with money"—that is, most places popular among the rich didn't typically value traditional wooden boat construction (too much maintenance!). And so they moved from a house on fifty acres to a former corn shed behind Dan Adams's brick barn, with winter approaching and work on *Rebecca* about to begin. David's first job was to hunt for the trees on Dan's land that would become the poles supporting the structure in which he expected to spend the next year and a half of his life, and then who knew how long after that.

✳

Nat Benjamin taught himself how to draw boats by reading, by working on boats, by sailing them, and by staring at them. When certain people stare at boats long enough, pretty soon they need to start drawing them. Nat had felt an affinity with boats ever since he built a raft as a boy to ride like a Hudson River Huck Finn, and when he stopped sailing as a way of life, he began to put the boats he had in his head down on paper. One thing about boats is that the very sight of them is infinitely engaging. You can look at a well-designed wooden boat for hours and hours over a period of years, even decades, and never tire of it, never see all there is to see, never fail to be elevated by the aesthetics of that one shape. Those curves, they draw you in, tantalize you, because they don't end; there's always a little more curve just out of view that keeps it interesting forever.

Rebecca began as an idea in Nat's mind for a full 60-foot schooner. He then shrank it down exactly, reducing each foot to ⅜ inch to create a profile sketch, and then enlarging it again to ½ inch per foot for the lines drawing, each curve precisely mapped out. On the lines drawing, the hull of the boat was divided into ten equal sections, or "stations," and from this scheme Nat created the diagonals, waterlines, buttock lines, body plan, and table of offsets (offsets are the measurements of the three dimensions of the hull: its height and width, where these heights and widths fall at any given point along the length of the boat, and diagonals, distances measured from the center line to sections drawn diagonally through the hull).

"I have known bright people," writes Bud McIntosh in his singular book *How to Build a Wooden Boat,* a 255-page manual between hard covers on how to build a 39-foot sailboat,

> to whom a lines drawing resemble[d] a cross section through a barrel of frozen angleworms, and meant but little more; and these same people thought of a table of offsets as something you might expect to come from the maw of a mad computer that had been fed on Pictish runes, rock and all. Both these conceptions are faulty and exaggerated. If you have managed (as I did, rather late in childhood) to master the technique of drawing a line from 1 to 2 and so on in proper sequence to 87, and got for your diligence the picture of a nice horsie, you should have no trouble with a table of offsets.

From this easy-as-a-horsie table of offsets, Nat made pencil marks on a big grid on the lofting floor of Mugwump and, by connecting those marks, reinflated the boat to its proper size. Drawing a boat on a piece of paper is simply a miniature version of doing it life size; just as ducks and splines had been used to create fair curves on the drawing board, nails and spruce battens, some as long as seventy feet, were used on the floor of the shop. The loftsman put nails at each X demarcating a certain line, a batten was held tight against this curved line of nails, and a pencil was dragged along the length of the batten. The pencil was then reinserted behind the ear—one of the defining characteristics of the boatbuilder here—or, in the case of Nat Benjamin, slid between ear and beret.

Nat liked to begin with perhaps the most visible line of a boat, the

sheer. And so at each station he measured the distance from the waterline (one of the main grid lines) up to the sheer, taking the number
from the table of offsets. At station five he measured 4 feet 1½ inches
up from the waterline; at station one where the bow is higher, he measured 5 feet 3 inches. Once he had the sheer plotted at each station, he
drove a nail into each mark, took a thick batten, one about 1¼ inches
square, bent that against the nails, and drew out the complete sheer
line. He and David put down on the floor each line of the boat this
way; the entire lofting process would take about a week.

For Nat, who has been doing this for more than two decades, lofting is fairly easy, but the transom is always difficult. *Rebecca*'s transom is
both curved from one side to the other and raking, meaning that it
leans back—two extra variables to contend with. Nat would offer an
article written by a notable Maine boatbuilder as proof that the process
can't be written about in words.

"I have an article by Arno Day about lofting the curved raking
transom," Nat said. "He tries to describe it over several pages, describes
several different methods, and I read it with some regularity. I still don't
understand it."

Considerable time is spent getting the transom exactly right. And
all along the loftsman is fairing as he goes, because there are many
places where one can err when enlarging a boat design to full size,
given that the width of a pencil line on a drawing translates into several
inches on the full-size drawing of the boat.

Nat's bangs were short, and the back of his hair curled out
from under his beret to the thick green hooded sweatshirt he wore in
winter. He typically donned the standard carpenter's canvas Carhartt
trousers, which began life vaguely orange and eventually faded to
vaguely mustard, by which time they were a relief map of hardened
bedding compound, dried glue, and red lead paint. Nat wore white
socks and ratty Top-Siders—"proper yachting shoes," he called them in
his throaty basso—or heavy workboots if the weather was very cold.
In addition to the pencil, he carried with him always an Opinel pocket-
knife, a folding rule, a bevel square, and a red bandanna on which to
blow his nose. If he was making patterns for the keel pieces, a hammer
would always be nearby, along with a bag of bright 4d nails in an Ace

Hardware brown paper bag—nails that would be used over and over and only part of the time pounded straight down; the other times they would be laid flat, and their heads pounded into the floor along this or that curved pencil line.

The low winter sunlight beamed straight in through the Mugwump doorway and across the lofting floor all morning. As Nat got to work, he gazed at the lines on the floor and then, when he had decided what the first order of business would be, cleared his throat loudly—*HhhhhHHHEH!*—and finished with "Aaaaaallright." The prelude completed, he dropped to his knees on the sun-golden lofting floor and held his stick rule across a piece of quarter-inch plywood on its way to becoming a pattern.

Patterns—exact outlines of the pieces of the boat—were the next critical stage of the process. "It's really just a big puzzle," Nat said of the wooden boat generally. "You draw the lines on the floor, you cut out pieces, and when you have enough pieces, you start putting them together."

Every shape of the boat's backbone was on the loft floor—keel, forekeel, stem, sternpost, knees, horn timber, and the rest; patterns of each piece were created by copying the shape on the floor onto a piece of plywood, or onto several pieces of plywood screwed together. Plywood is not transparent, so you can't simply take a piece, lay it over the shape you want, and cut it out. This creates the problem of how to duplicate the shape on the floor onto plywood with mirror-perfect accuracy.

A common method, and the one Nat chose for the sharp curves of the rudder, was to use box nails laid sideways. If a builder wants a half-moon shape, for instance, he lays a nail flat on the floor every few inches or so and pounds the head of it into the line of the curve, so that half of the nail head protrudes. Plywood is then placed on top and pounded on, the nail heads thus making a series of indentations in the shape of the curve underneath. The boatbuilder then turns the plywood over, pounds a nail conventionally into each indentation, holds a batten against these nails, retrieves a pencil from behind his ear, and drags it against the batten to reproduce exactly the curve represented on the floor.

Nat and David's main tactic, though, was to use what Nat calls jogged fingers. These fingers, each about 18 inches long, have a section removed so that plywood can be slipped underneath them. A series of them are nailed down with one end on the line being transferred, the opposite end of the fingers creating a duplicate of the line. The builder then slips a piece of plywood underneath the jogged fingers, holds a batten against them, and draws the curve.

With either method, the pattern on the plywood is then cut out with a handheld jigsaw, or on a band saw, to reveal a silhouette of the desired piece of boat.

(In a boatshop, such plywood patterns lie all over the place. Often they are dismantled and tossed in the stove. Sometimes, for *Rebecca,* they were so long and wobbly it took three people to carry them, one person supporting the middle. Compared with the mighty timbers of angelique that they gave their shape to, the ¼-inch ply patterns seemed flimsy as paper. They looked like useless scraps. Step on a pattern for a keel piece by accident, though, hear the wood crack, and Nat would come about as close to raising his voice as he ever did: "Don't break it!" he'd say. "That's the most important part of the boat!" But he never lost his equilibrium for long. He'd find a small piece of wood to screw into the pattern above the crease of the crack, a plywood bandage, and say, "If it's wood, you can fix it.")

The pattern is then held against the big timber, an outline is drawn, and the actual piece is cut with a chain saw or Skilsaw.

With the Mugwump pole shed built, the earthen floor covered with plywood painted white and now dynamic with the pencil lines of a schooner, and numerous patterns lying about, David and Nat rolled the main keel timber, the first piece of the boat, into the shed. It measured more than half a foot thick, 2 feet wide, and 35 feet long. Nat and David did not have a forklift or other heavy machinery to move it, nor did they have a motorized block and tackle overhead, as Brad had had in the lumberyards of Suriname. But they did have a small hydraulic jack, some heavy metal poles to use as levers, and metal and wood cylinders that served as rollers. These items were more than sufficient to enable the two of them to maneuver a one-ton main keel timber and a three-ton forekeel. They simply levered the wood onto rollers and then levered it forward, the weight creeping along not unlike the stones of the great pyramids in Egypt four thousand years ear-

lier, David or Nat racing the back roller around to the front of the timber as it progressed (Nat invariably referred to G&B's heavy-lifting
technique as "Egyptian technology"). Raising these pieces to different
heights was often just a matter of balance: one person could lift one
end of a 6,000-pound piece of wood a foot or more if it was balanced
properly. "It's amazing what you can do with rollers and levers and
jacks," David said.

The first piece of wood cut for *Rebecca,* the main keel timber, did
not require a complete pattern and was not fully drawn on the lofting
floor. When Nat designed the boat, he believed that the best width for
Rebecca's keel would be about two feet, but he didn't know if Brad
could get him something that wide. In fact, the timber Brad ultimately
sent for this piece was about an inch shy of the ideal width, but it com-
pensated for the deficiency with its virtual lack of knots or checks.
Furthermore, there was no way of knowing ahead of time if there
would be sapwood—the soft outer layer, typically a map of worm
trails—intruding on the piece of wood. So when Nat and David got the
timber into the shop and up on blocks in the center of the lofting floor,
over its pencil image (what there was of it), Nat knew he would have to
let the piece of wood tell him how wide to make the keel. The eventual
shape was drawn onto the timber to maximize the width, cut out, and
then transferred onto the lofting floor. It was a fairly simple, rectangular
shape—35 feet by 2 feet by 7 inches, tapered toward each end—and
would for all its life lie bolted against the 26,000-pound lead ballast.

The next piece of *Rebecca*—the forekeel—was a little trickier than
the straight keel timber. Nat had designed *Rebecca* with a cutaway forefoot, meaning that instead of having a bow that dropped straight down
and connected almost perpendicularly to the keel, like a plow, her hull
would slope from stem to keel at something closer to a 45-degree
angle, more like a bullet, via the forekeel piece Brad had been so lucky
to find—the one with the long, lazy curve. This forekeel was huge.
More imposing than its weight, though, was its 1½-foot width. The
only way you could cut this monster timber was with a chain saw, but
it would be almost impossible to guide a chain saw accurately through
so much angelique. And this piece of wood was too remarkable, too
important to *Rebecca,* to allow for any errors.

Ross remembered seeing a colleague—Gary Maynard, who ran

the *Alabama* project, the restoration of a 90-foot schooner originally launched in 1926, a project then nearing completion across the sound, in Fairhaven—rough out big spars by rigging up a two-handled chain saw, and he figured that would work here. So he drilled two holes in the bar a few inches from the tip of one of the yard's chain saws—no easy task, since it was solid steel and not intended to be drilled. He then fabricated from scrap bronze an attachment to fit through these holes, bolted it to the chain saw, and attached a 2½-foot-long oak handle to the bronze fabrication, long enough to give the man on that end as much leverage as the man holding the motor end. Nat and David then carried the contraption over to the forekeel timber—the pattern had been drawn on both sides, and a center line drawn across the top—and began ripping through the wood. They started a couple of inches above the actual line, not knowing how accurate the gizmo would be, but they, and curious others who tried it out, soon found it to be so accurate that they could approach the line up to ¼ inch.

Having cut the rough shape of the backbone timbers, Nat and David got to work with Skilsaws, broadaxes, and adzes to reduce the pieces of wood further and get them nearer to the shapes of the patterns. When they approached the shapes, they used power planers and then eventually hand planes, until the timbers were exact. They cut scarfs on both ends, long tapered joints where the forekeel would fit into the keel timber and into the stem, the keel to sternpost, horn timber to transom knee. And finally they cut the rabbet.

One of the most critical parts of the boat, the rabbet was not an actual piece but rather a vacancy, a V-shaped groove chiseled into the backbone where the planking would land. Nat and David marked the rabbet onto each backbone piece, just as they would any other pattern, and cut it out carefully, bolting the backbone pieces together after the rabbet was mostly cut.

Nat and David maneuvered the forekeel, which now weighed considerably less than its original 3,500 pounds, to the keel timber and balanced it on a fulcrum so they could easily lift it to adjust its fit by planing the surfaces that would be connected. If the pieces didn't rest perfectly flush, the keel would be weaker, and water could collect in the gaps and begin to rot the wood. "You want it to be onionskin," Ross said of the fit of these keel pieces.

When the fit was perfect, the surfaces were slathered with tarlike roofing cement and bolted together with ¾-inch-thick bronze rods, sporting threads cut on the Hendey metal lathe; the bolts were countersunk on the outside of the timber, eventually to be filled with an angelique plug so there would be no holes.

Thus the backbone of the boat took shape, piece by enormous, hand-worked piece. It happened virtually the same way for big *Rebecca* as it had for smaller boats such as G&B's 44-foot schooner *Lana & Harley,* or *Blue Rhythm,* the 21-foot gaff sloop the boatshop had launched the previous summer. The pieces were the same, but with a small sailboat you didn't have to move them around using levers and rollers. A few months into the project, two new crew members arrived, Pat Cassidy and Todd McGee, who had been working on the *Alabama* project under Gary Maynard. The 150-ton *Alabama* was to *Rebecca* what *Rebecca* was to the 44-foot *Lana & Harley,* so these young men were used to working with truly large timbers, 3-inch-thick planks, deck beams that spanned a 21-foot width. When they first arrived at Mugwump, Todd and Pat kept the others' awe in check by noting how pleasantly tiny *Rebecca* was, by bolting in yet another "dinky" deck beam. *Rebecca* was almost like a toy to them.

"To me this is a big boat," David said. "I'm used to small boats." But *Rebecca,* he continued, "is still really what we call a small boat. Almost any piece can be handled by a couple of men instead of a forklift. You get into bigger schooners, you're using forklifts and come-alongs and jacks, and you don't even consider moving things around."

Regardless of your perspective, whether you were straight off *Alabama* or came from a small, one-man shop in Maine, *Rebecca* had a powerful magnitude. It was by far G&B's biggest boat. *Lana & Harley* (now called *Calabash*) was 16 feet shorter than *Rebecca.* But length was not the true measure of a boat. *Lana & Harley* weighed 30,000 pounds; the 60-foot *Rebecca,* just 36 percent longer, would be 156 percent heavier, at nearly 77,000 pounds.

That was big enough for Nat and Ross to decide to ask a naval architect to perform some calculations based on *Rebecca*'s lines. For all their boats thus far, Nat and Ross had estimated the amount of ballast that would be needed according to standard ratios. And they had enough experience sailing boats of the sizes they built to get it right.

They had misjudged slightly on *Lana & Harley:* after being launched, she'd sat higher on her marks than Nat liked, and both Nat and Ross had felt a little tender under sail, so they'd added an extra thousand pounds of trim ballast, and now she sailed beautifully. Nat liked to ballast a boat slightly light; he'd seen enough boats that were ten or twenty years old riding low on their marks because of extra equipment their owners had loaded on after construction. *Rebecca* was a much bigger boat by far than they'd built before, and so some precautions were warranted. Also, Nat was curious to know how modern techniques and computerized readouts would compare with his eye and his "ancient arithmetic." So it was worth the small expense (to Dan) to hire a naval architect, especially since that architect was Ross's nephew Antonio Salguero, his older sister's son, who'd practically grown up at the yard and whom Nat had watched develop. Antonio had studied yacht design and had begun his own business as a boatwright and designer in Port Townsend, Washington, the West Coast center of wooden boats. "I really respect his judgment and expertise," Nat said of Antonio.

Antonio, not yet thirty at the time, was thrilled and honored to be asked to do the work on what he felt might be, in his words, "the pinnacle of Nat's career." It would also give him a chance to put to use what he'd learned at the Maine Maritime Academy in Castine. And so he created a construction drawing and ran some numbers to determine such things as optimal engine size, skin-friction coefficients, wave-making resistance in pounds, and stability curves, as well as, of course, how much ballast *Rebecca* would need and where exactly it should go on the boat, given her center of gravity and her center of buoyancy.

To decide on the weight of the ballast, and to make the boat's waterline fall exactly where Nat had drawn it, Antonio first had to calculate how much water *Rebecca*'s hull would displace, which gave him a volume of water and therefore its required weight. He then added up the weights of every piece of the boat—every keel piece, every frame and deck beam, all the wood, all the rigging, all the bolts, all the hardware that would go on, the weight of the anchor and its chain—and the weight of the fire extinguishers, the life jackets, the linens, the galley stove, every single thing on the boat insofar as anyone could know such things exactly (Antonio noted that Nat liked to figure out a boat as he went, leaving many decisions up to the builder at the moment of construction: "That's the way Nat does things," Antonio said. "Very organic planning"). The boat was traditional enough in its design to allow good guesses, and after plugging in all the numbers, he determined that the boat and all its contents would weigh 50,015 pounds. From Nat's drawings he knew that the boat needed to displace 76,867 pounds of water, which meant that a ballast of 26,852 pounds should be cast—almost 35 percent of the total weight, a good ballast ratio. Despite the exactitude of the numbers, much was guesswork, and Nat furthermore wanted the boat to carry about 2,000 pounds in trim ballast that could be jettisoned should the owner load up the boat as the years passed. Nat was especially gratified to learn that Antonio's figures, gleaned from complex computer programs, concurred completely with the figures he'd arrived at using a pencil.

✳

Antonio would also be tapped to build *Rebecca*'s two masts for a marconi mainsail and a gaff foresail. The marconi rig—tall mast, triangular sail—is all but ubiquitous in the industry; the gaff rig—four-sided sail attached to the boom along the foot of the sail, shorter gaff above attached to the head—is all but nonexistent. The gaff rig vanished from the mainstream seventy years ago. Boats built before the 1930s were designed to carry gaff rigs, and so they can still be seen here and there—typically at classic boat shows and regattas—but a gaff rig on a new boat is an oddity. That part of *Rebecca*'s rig was to be gaff, though, was hardly unusual for a G&B boat. What *was* unusual in this Benjamin design—unprecedented, in fact—was the marconi main. Most of Nat's boats had been designed with a gaff rig. And furthermore, no one here seemed to think that the least bit peculiar or remarkable.

A few people have written about the gaff rig, with varying measures of appreciation. No one argues that it's not beautiful to look at. As Ross put it, "When you see the silhouette of one, or you're on one and you look up, it takes you to another time." Ross moreover maintained that such rigs were easier to sail and, because they put less strain on a boat, better for the boat. But historians of the rig have noted, for instance, that by "the 1950s anyone ordering a new yacht with a gaff rig would have been considered eccentric." And by 1975 "the eclipse would become total . . . gaff rig sailors were an extinct breed."

Gaff rigs came into use in the 1600s and quickly proved to be an efficient and effective design. Historically, the first sails had been rectangular or square, a shape that was used almost exclusively for at least four thousand years. Such sails still work quite well today for those few ships that carry them, but they're no good if the wind is blowing straight at you. A rig running fore-and-aft, as opposed to athwartships, was developed precisely to surmount this problem, and sails grew progressively more triangular. In the 1800s the gaff rig evolved rapidly because of large-scale manufacturing of equipment such as sails, blocks, and lines; improved ironwork and the development of iron wire rope allowed bigger ships to carry bigger rigs and more sail. The rig predominated until the 1920s, when the marconi or Bermuda rig, which had been used in the Caribbean since the late 1700s, emerged as a popular new choice, one that many people thought would be a passing fad. But workboats soon traded rigging for engines, and yacht design, always shaped in large measure by the quest for speed, found that the marconi was faster to windward and could point closer to the wind by far than the gaff rig. And how fast a boat was to windward was always the ultimate definition of a boat's speed, at least as far as selling rigs was concerned. The gaff rig's numerous advantages were ignored. It was more powerful than the marconi, many said, when the wind moved more than 60 degrees off the bow; a variety of sails could be set, and the shape of the sail could be fine-tuned by adjusting the peak or the throat of the gaff; a gaff rig could carry a lot of sail, making it perfect for heavy boats; it was easier to handle; the mast was shorter and therefore easier overall; and the rig strained the hull less. But never mind all that—how fast was it to windward?

Nat was never willfully retrograde—like those Creative Anachronism guys who re-create Civil War battles—or wistfully nostalgic. Going fast to windward or beating hard into the wind was never his main concern. But if you wanted to talk about speed, he'd tell you that gaffs *were* faster than marconi rigs downwind—on many points of sail, for that matter—and furthermore didn't require all kinds of sail changes to get the boat to do what you wanted it to do. The gaff rig was simply the best rig, as far as he was concerned.

"For cruising boats, you don't go hard on the wind very often," Nat said. "For a cruising boat, a family boat, you don't need high performance. You want something comfortable and simple. It's very simple. Everything is held together with pieces of string; there's no fancy outhaul, worm-gear castings, cunninghams—it just doesn't need that stuff. Very low-tech. Most people really aren't into high-performance racing. Most people just want to go out there with their family, have a picnic, and enjoy themselves. Most people want simplicity and comfort and safety. But the reputation that gaff rigs are slow is false.

"They're old-fashioned, they're not marketed real well," he continued, reflecting on the gaff rigs' virtual extinction. "There were few around here till we started building 'em.

"We sailed *Zorra*—she was a fabulous boat, a big, powerful seventy-two-foot boat—sailed her up and down the Caribbean and around here, all over the place, and had spectacular sailing on that boat. But the luff of the jib was eighty-nine feet. That's a long luff, and it's a one-to-one halyard, there's no purchase, so it's a lot of work. It's fun, it's very invigorating. But sometimes it's nice to have something a little simpler.

"And it's easier from a builder's standpoint. There's a lot less fabrication, a lot less hardware. It's more wholesome to me."

Thus Antonio would fashion a foremast almost 59 feet tall and 10 inches in diameter at its widest point, from deck level on up to the highest reach of the gaff jaws, after which it would taper to about 6 inches. Nat opted to give the mainmast a marconi rig, though, because while a gaff rig can be simpler, a gaff-rigged mainsail in a boat that size can be more labor-intensive than a marconi sail; Nat felt that a marconi mainsail would better suit the buyer, and furthermore, he had enjoyed the similarly rigged *When and If.* The 72-foot mainmast would have a uniform taper. Both would be made of Sitka spruce from Canada.

✳

Rebecca, while not enormous by yacht standards, of course, was nonetheless one of the biggest new plank-on-frame boats under construction in the last half of the century, big enough to cross any ocean and make the lives of the many on board comfortable and safe on virtually any crossing. And more, she would look every bit her size. A two-masted traditional wooden vessel 60 feet long looks majestic at anchor in any harbor, but heeled over, bombing across the ocean, her 1,800 square feet of sail set tight, *Rebecca* would look awesome.

"Often you're working on a boat that someone *else* thinks is beautiful," said Todd McGee. "This is a beautiful boat. It's got powerful lines up forward, but back aft she's very graceful, elegant." It was that combination of power and grace that many people working on her noted with pleasure.

The gradual accumulation of pieces, this majesty and awe in the making, were on view every day as the backbone took shape during the month of February 1998. When the backbone was complete, double-sawn frames of white oak—the boat's ribs—were bolted together; patterns were drawn on them, and they were cut out on the ship's saw, whose tilting blade beveled their edges to whatever predetermined degree would accommodate the planks that would be bent around them and bolted in flush. As these frames piled up, Nat created a plug for the lead ballast, which was poured at a foundry in Providence, Rhode Island, delivered to Mugwump on a flatbed truck, maneuvered to the center of the building, and attached to the keel timber with custom-made 1¼-inch silicon-bronze hanger bolts.

Two layers of ⅝-inch silverballi were steam-bent over a form, left to dry for a week, and then glued together with epoxy to create the curved, laminated transom.

And soon the frames began to go up, bolted to keel and floor timbers, then faired. *Fairing* and *fair* are broadly used terms that refer to a boat's curves. You fair a curve, whether a single frame or an entire hull. The sheer on a beautiful boat is a fair curve. Scrutinizing the line of a plank, one eye closed, a boatwright might see an imperfection, a hump or a dip in the line, and say, "That plank's not fair." (If Nat was in earshot, he'd say, "*Life's* not fair!") And the big heavy sawn frames, they needed to be faired so that the planks, as they bent their way around the boat, would lie like onionskin against their bevel. Fairing these frames was an arduous, painstaking task. Each fat frame, rough-sawn on the ship's saw, had to be carved with both block planes and spoke shaves till it was perfectly smooth, its bevel just right. Each frame of the hull had a slightly different bevel.

All this work would take the crew five months—Nat and David

and then Pat and Todd; an apprentice named Casson Kennedy; Brad
Ives; Daniel Feinstein, a man with more sailing experience (including
two Atlantic crossings in his 25-foot double-ender) than his age,
twenty-three, would seem to accommodate; and Mark LaPlume, an
itinerant woodworker, bongo maker, and visual artist. With all the
frames slowly being raised and bolted in, and then hand-planed for
many weeks, you couldn't help but stop several times a day just to stare
at this thing, this enormous wooden boat, its hull coming together before your eyes. Once all the sawn frames were up, you could for the
first time understand the size and shape of the boat, sense its scope.
The sight of it could halt you if you didn't know it was coming. Having heard about the schooner, passersby—and they were many—often
said they felt their breath leave them upon beholding this creature.

."There are sizable vessels being built here and there out of wood,"
David said, standing at the starboard sheer, on staging that had been
erected around the circumference of *Rebecca*. "But I don't know of any
that are really of this era. There are reproductions of older working
vessels, but I don't think there's anything being built to this scale that
would be called a yacht, that's all traditional. There are big yachts being
built, but they're all being built out of . . . goop." Fiberglass. "So what's
important is that this boat is being built to outlast by far any other boat
being made today—cold-molded, fiberglass, steel. It'll be here sailing
when we're long gone."

Daniel Feinstein, who intended to leave for England the following
spring under sail, raised a hammer high and brought it crashing down
on a fat bronze bolt, securing one deck beam after another. "They're
the only thing that man can make," he said of wooden boats generally,
"that's almost as beautiful as nature."

Daniel looked the part of seafarer, with his thick, dark, jutting
beard and a solid black bar of an eyebrow running across his eyes; from
way up on *Rebecca*'s sheer, hammer in hand, he shouted down to Nat,
"Some guys from the lumberyard came by to have a look yesterday." (It
is an irony of fate that each day, to get to Mugwump, Nat had to walk
through the Hinkley lumberyard, striding down an aisle of rectangular
stacks of boards—"worthless wood," he said, noting that lumber com-
panies accelerated the growth of the trees to such an extent that *hardwood* wasn't even hard anymore.) "All they wanted to know is how we

bent all those boards," Daniel continued. " 'How did you bend the wood?' they kept asking. That was all they wanted to know. They don't understand anything that's not precut into two-by-fours." He laughed, then brought the mallet down hard onto a bronze bolt, driving it into the sweeping oak deck beam.

✳

Once the double-sawn frames were faired, planks of angelique nearly 2 inches thick and half a foot wide—bilge stringers—were bent around the lower interior of the hull, and oak sheer clamps were bolted along the top of the sheer. Then the whole hull was covered on the outside with 2-by-3-inch spruce ribbands up to 20 feet long apiece. These served as a temporary mold against which the boatwrights would bend three white-oak frames between each sawn frame, running from keel to sheer. To bend this 2½-inch-thick white oak without breaking it, they cooked it in a steam box for hours till it was soft. Ross had built one steamer out of some old propane tanks, using a cutting torch and welder; it was simply a big pipe with a tank of water attached to it below. The current G&B model was a long wooden box. The Mugwump crew made theirs out of a section of corrugated-plastic culvert pipe. The water was boiled over a propane flame, and once the steam "box" was hot, in went the frames, and then a wooden door was wedged tight against the opening. Not only did steaming make this wood pliable; many believed that it also helped to preserve and season wood by killing bacteria that caused rot. When they were cooked, the frames were taken damp and steaming out of the box and immediately clamped fast against the ribbands, then secured with screws. After the fifty or so frames were in—between the stations where the masts would be, Nat had used all double-sawn frames for extra strength where the rig stressed the hull most—*Rebecca* was ready to be planked.

People often use the words *nature* and *natural,* as Daniel had done, in their attempts to express their reverence for a wooden boat. Part of a boat's connection to nature, obviously, is that its shape mirrors those found in the sea, in large fish and limbless mammals, forms largely defined by long, smooth curves. Fish aren't composed of a lot of right angles, and neither are sailboats. Curves make a sailboat. Often those curves are sawn out, as with the heavy white oak frames bolted in every four and a half feet along *Rebecca*'s keel. Large deck beams have some arc to them, are often chosen with some sweep, and must be sawn out. But other curves are made by bending wood, such as the steam-bent white oak frames. And just as Nat, on a stool at his board, had bent thin plastic battens to draw the curves of a boat on drawing paper, just as spruce battens had been bent against nails to transfer those curves from paper to the lofting floor, so long pieces of wood—planks—were now bent around sawn frames to form the hull, and screwed into the intermediate steam-bent frames. Because a boat is largely defined by her curves, the act of making these curves without a saw, bending the curves into the wood, is somehow more exciting than almost all other tasks for a boatwright. There is tension in a curve—sometimes an oak frame wrestled into place against the ribbands breaks with a heart-jolting *crack!* And a curve made by bending clear stock is almost always true.

Fiberglass boats often have true curves, but that isn't a given. You can make a fiberglass boat in any shape you want, unlike a traditional wooden boat. Thus many designers *do* make them any shape, most often enlarging the belly unnaturally to create more sleeping space below. The shape of a traditional wooden boat is limited by how far wood can bend.

Planking again is exciting work because it is so big and visually dramatic—if there's one thing that makes a traditional wooden boat what it is, it's the planks. All planks, some of which exceed thirty feet, are first planed to a 2-inch thickness and then spiled, the term used for measuring and drawing their shape on the board. One of the marvels to nonboatbuilding boat lovers is how all these miraculous planks are made so that they fit together, given that a single plank is very fat in the middle, bent from end to end, where it might narrow to a point, and often curves up at one end and down at the other like a lazy S.

There are a couple of ways to line off the planking, but generally it's a matter of dividing the hull into three sections and then dividing each section by the number of planks it will carry. The biggest frame, where the boat is fattest, is measured first to determine how many rows of planks the hull will have. David, in charge of the bottom section, knew he wanted fairly narrow planks, 7 inches at their widest point. He divided the length of the biggest frame by 7, and that was how many rows of planks he would need; they would use that number to divide each frame section. A plank that was 7 inches wide in the middle of the boat would narrow to 3 inches as it approached the bow and then to a thin, tapered end by the time it reached the stem. To get the desired plank shape, David would mark off the widths at each station onto a spiling batten, then transfer the batten measurements onto the silverballi plank and saw it out with a Skilsaw. Each plank would need to be adjusted with a hand plane and fitted exactly, and a caulking bevel would be planed on one edge to allow a space for the cotton caulking to be wedged in. David himself would measure and cut out the plank, and then a crew behind him would do the final fitting, hanging, and fastening (when a plank fitted perfectly, it would be clamped in place, holes would be drilled, and bronze screws would be countersunk into the plank, the holes eventually to be filled with silverballi plugs or bungs). The teams flew through the work in less than two months; a shutterplank party was held at the end of September to celebrate the hanging of the final plank and the closing of the hull.

The hull, at this point, was still composed of numerous though scarcely visible right angles—the uneven edges of the planks—and so it was faired, made perfectly rounded and smooth, with electric and hand planes, power sanders, sandpaper affixed to flexible boards with handles (called "torture boards"), and glass-foam fairing blocks. Fairing took the final edges off the hull.

The last step in making *Rebecca*'s hull watertight was caulking—filling the space between the planks with strands of cotton, using steel wedges, called caulking irons, and hammering with wood mallets. Two strands of cotton were used in all seams but the lower five, which first got a strand of oakum—hemp fibers treated with tar—that would withstand the corrosion of diesel fuel should any be spilled in the bilge. A professional caulker named Frank Rapoza lived on the Vineyard, and Nat hired him to begin the work and give the crew lessons; most of the crew had caulked before, but here they learned from someone who was a caulker by trade. Any boatwright passing within fifty yards would know from the sound what was happening inside Mugwump. Wood mallets striking iron striking cotton reverberate in the 60-foot drum with a distinctive *thunk*. With half a dozen caulkers at work, a "funky rhythm," as David described it, sounded out all day long, muffled and pleasing, the drumbeats of a demented reggae band. Happy music—a boat was being sealed up, this was going to happen!

When all the seams had been caulked, they were puttied. The hull was now watertight and would float. She would soon be painted with lead paint—called red lead, though it's actually orange—a primer for the copper bottom paint used to deter teredos and other wood-boring water bugs that might eat away at the hull.

✸

At the end of September, as the planking neared completion, Ross took *When and If* up to Maine for a short break from the boatyard, a trip that would forever alter his life, though to what extent he would not know for more than a year. September on Martha's Vineyard— after the frenetic work of summer, after the summer throngs had departed, when the island, no longer burdened, sat a little higher on her marks—was a month that featured various local races. At G&B September was informally considered sailing month. While Nat and his crew had been blazing away on *Rebecca,* Ross had been running the boatyard, building dinghies, repairing tired old boats, dealing with summer maintenance, and constructing another Bella, this one for Pulitzer Prize-winning writer, historian, and TV host David McCullough, who lived on the island. All of this was valuable work, but not very exciting relative to what was going on up the street at Mugwump.

So it was good for Ross to get away for a few days, to sail, and to see friends down in Maine, many of whom gathered to watch Donald Tofias launch the second W-class racing sloop, the 76-foot *White Wings,* built by Taylor Allen at Rockport Marine, a couple of miles south of Camden, Maine.

It was here that I first met Nat and Ross. As Ross and I sat below on *When and If,* he described the repair of the impressive schooner that had nearly been scrapped after being driven up on rocks in Manchester Harbor in 1990. A publishing executive named Jim Mairs had bought her in partnership with Nat and Ross, who had submitted the lowest bid for her repair. They would come in even under that, Mairs said, probably because, unlike the proprietors of most of the other yards vying for the work, they understood the nature of the damage and the strength of the existing hull.

"You can fix a wooden boat," Ross told me that afternoon, seated below deck in the saloon, leaning against mahogany, the low afternoon sun shooting through the companionway to light up motes of dust in the air and glare off the varnish. "They're forever repairable." All day he showed visitors the new port-side planks, visible from inside the cabin, that he and Nat had put in, replacing the entire middle section of the boat's hull.

Ross lamented with a kind of smoldering disdain everything that was cheaply and quickly made. "Even *hardware* isn't hard anymore," he said. It was plastic, everything as weightless as could be. Tools today were no good, they didn't work. The older the tool, the better, as far as he was concerned. By "old," he meant century-old. He collected these items for the Gannon & Benjamin boatyard. He scavenged old sea towns for antique hardware to install on the yard's boats. Not for authenticity; these fittings were usually superior to anything made within the last thirty years, he explained. And usually nobody wanted them anymore, so you could get the stuff cheap, if not for free. This practice of using strong, antique hardware underscored the fact that while there was great aesthetic pleasure to be had in building and sailing wooden boats, it was not fundamentally an artistic endeavor—Nat and Ross weren't artisan builders. They were sailors.

✳

The plan was for Ross to sail *When and If* up and anchor at Rockport on the day of the launch; there he would meet Nat and Pam, who had driven up, and they would trade vehicle for vessel for the return trip home. Whispers of Nat's arrival that day preceded the man. *Where's Nat? Have you seen him? I heard he was here. Yeah, I heard he was here, too. We haven't seen him, either.* But by midday the corporeal Benjamin appeared on the dock with his wife, Pam, tall and beautiful with her thick blond hair going to gray, friends surrounding them. They would take the boat today, do some cruising along the coast, then head back to the Vineyard. It was the kind of sailing that Nat rarely did anymore, and he missed it. When he and Pam and their children were under sail, he felt at home, his senses fully engaged. Nat wrote to friends after this trip, and evident in his recollection is the energy, a kind of nineteenth-century grandeur, that infused his spirit when the schooner was under way:

From the log: September 28, 1998. 1030. Moored just north of Birch Cove, Bartlett's Island . . . we slip our mooring and under main, fore, and forestaysail, ease off on a port tack to beat up through "The Narrows" between Bartlett and the west side of Mt. Desert Island. A freshening northwest wind lifts us clear of these noble straits into the sparkling inland sea of Blue Hill Bay. Sheets eased a bit and her deck hissing against the press of vanishing flat water, When and If *carries us beyond our expectations into that unique dimension of life under sail. All hands find their right place—out on the bowsprit, up in the rig, in the lee of the doghouse, at the helm—in a comfortable niche to absorb the visual feast of gradually changing vistas from the distant Camden Hills and Blue Hill to the west and north, Isle au Haut to the southwest and Mt. Desert to the east. These majestic landmarks provide the background for the moving presence of closer, lower islands with granite fringes and tall spruces spearing the sky—an evolving scenery choreographed by our silent stately advance.*

Midday arrives and lunch appears on deck—time to run out the sheets and broad-reach down the west side of the lovely Bartlett observing coves, headlands and the peaceful green pastures which nourish prize Angus and Siementhal cattle wandering about on their private island heaven. Eventually we fetch the southern point off Ram Island and must pull ourselves out of reverie and digestive cycles and trim all four lowers to face what has now become a gentle afternoon breeze. The land closes in on us as we maneuver between the verdant shoreline and curvacious outcroppings of earth-toned rocks. Sails aloft draw on zephyrs unnoticed seventy feet below and pull When and If *through the narrow cuts in shifting air with only an occasional nudge from the helmsman. Past another steep bluff and the anchorage comes into view in softer afternoon light. Now clear of the wind shadows, clean air slips by 1,800 square feet of sail cloth and* When and If *accelerates home like the horses, gazing at us from the fields above, head to the barn at the end of their day. Headsails come down and we shoot 43 tons on a long, measured approach to the mooring. Made fast, all hands move about on a thin cushion of air, lowering sails, stowing gear. . . .*

IV

Autumn lingered on the island, so that even at the end of October, with *Rebecca* nearing the halfway mark—her hull planked, caulked, faired, puttied, and painted below the waterline with red lead, and half her deck beams in—the days were warm enough that Nat needed to pull on only a couple of shirts and a wool sweater in the morning. He headed downstairs, ate a piece of toast sweetened with honey kept in an enormous jar and a bowl of cut fruit, then grabbed drawings of *Rebecca* that he'd worked on over the weekend, stuck them in his canvas shopping bag—his briefcase, he called it—and headed out the back door. This part of Martha's Vineyard was thoroughly suburban, with tall trees and well-kept lawns. Nat crossed his own backyard and cut through his neighbor's, strode up a dirt drive strewn with dried leaves from the surrounding white oaks, passed a small graveyard, then turned left on Main Street toward the town of Vineyard Haven. Nat has a distinctive walking style recognizable from a long way off: he bobs in a slow, hulking manner, his feet spread wide, legs almost bowed, and moves at an impressive though effortless clip. Remarking on this walk, a friend of Nat's once laughed with appreciation and said, "He walks as if he'd spent his whole life at sea." When Nat arrived at Beach Road, he turned left toward the water, passed the Black Dog Restaurant, and crossed the sand to the weathered G&B boatshop.

Duane Case was at work on one of two Bellas nearing completion. Jim Bresson, a young Frenchman, would arrive soon for work, as would Bob Osleeb and other regular G&B crew members. Nat had a few words with Duane. Doug Cabral, editor of the *Martha's Vineyard Times,* had stopped by on Saturday to tell Nat that the boom on *Liberty,* his 40-foot G&B-built sloop, had snapped in half during a sail the day before; this irregular grain would need to be spliced, Nat explained to Duane. Nat paused in the office to speak with Ginny, then headed across the street, through the Ace Hardware parking lot, past stacks of Hinkley lumber, to Mugwump. Ted Okie had returned from vacation, ready to begin work, and Nat said, "He's *back*. You missed the excitement."

"I missed the red lead!" Ted exclaimed. The previous spring Ted, twenty-four years old and just out of New York University with a degree in urban planning, had been fishing around on the Internet for some kind of gainful employment when he spotted on the *WoodenBoat* home page a notice advertising slave wages in exchange for work on a 60-foot plank-on-frame schooner. He called Gannon & Benjamin, had a conversation with the ornery woman who answered the phone (Ginny), was given vague cautionary information, decided to show up anyway, and was soon put to work on *Rebecca.*

Nat entered Mugwump, walked up the plank ramp, climbed a small ladder built into the outside wall of the office, stepped onto the staging that circumscribed *Rebecca,* and headed around the transom to the port beam, where Pat Cassidy was managing the installation of the deck beams. Pat looked nervous when he saw Nat with the drawings, but Nat chuckled and said, "I didn't change anything," as he unrolled them. He and Pat discussed progress and today's work. Not everything was marked on the drawings. Pat couldn't, for example, check them to see where to put the block for a running backstay; he'd ask Nat where he wanted it, and Nat would look at the stock, the boat, the drawings, and make the decision at the moment. This was the nature of organic planning. Today Pat would continue to saw out and install deck beams with Daniel; Mark would bolt in cabin sole supports and begin work on the sole there. And Nat and David would head out to Dan's, where much of Brad's wood was stored.

With the framing of the deck moving so quickly, Nat figured it was time to get the decking down to Mugwump in preparation to be sawn out. Ross's trailer, which the yard used to haul wood and small boats, was hooked to the back of David's red pickup, which Nat drove to Dan's land. Stacks of Suriname woods were "stickered" near a lichen-covered stone wall—stacked resting on thin sticks of scrap wood to promote drying on all sides of the boards—and Nat and David spent half an hour flipping over heavy planks, looking for Rebecca's deck. They would examine one side of a board, staring at it for a while, and then, both of them grunting to push the planks over and jumping back to avoid crushing a toe, they would stare at the other side for a while, saying, "Oh, very nice," or "Look at that sway," or, when a plank had a crack through it, or its grain swirled like a weather-map hurricane, "Oops!"

Nat retrieved three steel pipes about 2 inches in diameter and a yard long from the back of the truck. "Good old Egyptian technology," he said. He backed up the trailer to the woodpile, and after they'd chosen a dozen planks, each 2 inches thick and some as long as 30 feet, they began to load them onto the trailer by first placing the pipes next to the board being loaded, then flipping the board over onto the pipes; once the board, weighing about 250 pounds, was on the pipes, it would roll with the gentlest push onto the trailer.

Nat stopped at Back Alley's in West Tisbury for a midmorning coffee before delivering the silverballi to Mugwump.

These planks would be stickered again as they waited to be sawn into strips 1¾ inches square, which would be laid across the deck beams, screwed down, and caulked. The very next steps after framing the deck were putting in the two teak cabin sides and beginning to build the interior.

Nat got to work upon their return, first sharpening the blade of his plane on a stone at the back of the shop. He would spend the rest of the day adjusting by eye and jack plane the starboard sheer—"one of the most critical lines on a boat," he noted—to make it perfectly fair. Two strangers who had heard about the schooner walked along the staging, having a look around. They were a middle-aged man and woman, one a pragmatist, the other a romantic. They hung over the transom and stared down into the cavernous hull.

"You can see the world in there!" the woman exclaimed.

The man said, "That's a lot of wood."

✴

That afternoon, after lunch, Nat walked back to the boatyard and out onto the dock. He hopped down into one of the yard's launches and untied her to give me a tour of the harbor and to check on G&B boats. Gannon & Benjamin had built many of the boats here, and the boatyard crew took care of even more, checking their moorings and adding chafing gear when storms approached; they were the de facto caretakers of all the wooden boats in this harbor that was filled with them all year round.

"This harbor is mind-boggling," Nat said, zigzagging slowly through the moorings of Vineyard Haven Harbor. "There's not another harbor like it anywhere."

Nat rarely praises things he's had a hand in, but this harbor was an exception. G&B was completing its eighteenth boat in as many years, all of them designed by Nat; more than half plied the Vineyard Sound (others had gone as far south as the Caribbean). G&B had also repaired or brought back from the edge of death a few dozen more that remained in this harbor. Moreover, Nat and Ross had created an environment that was favorable to wood, that nourished wooden boats and informed their owners. Without G&B—a yard that could maintain and repair traditional wooden boats—many people who enjoyed wooden boats wouldn't be able conveniently to own or keep them here because maintenance and repair would not be available. Combine Nat and Ross, and their work, with two other extraordinary wooden boat figures, beachfront neighbors Bob Douglas, owner of *Shenandoah* and *Alabama* (not to mention the Black Dog restaurants and catalog store), and sailor-shipwright Gary Maynard, who had restored *Alabama* and for years headed the Five Corners Shipyard, and you truly did have a harbor that was unique in terms of wooden boats.

(Anyone living on the other side of the country would want to trumpet Port Townsend, Washington, another wooden boat center that annually sees the arrival of hundreds of classic boats for its wooden boat festival. Scores live year-round between the port of Port Townsend and Point Hudson. Numerous workboats—seiners, longliners, halibut schooners, and tenders—still ply the waters nearby and make this a very active harbor in the repair and maintenance of traditional wooden boats.)

As Nat steered the launch, he nodded toward vessels on either side of his own boat, *Venture,* a 37-foot gaff sloop built in 1910. There was a New York 30 built in 1905, he said, and that little schooner was built in the 1920s; he pointed out two Malabars, some catboats, yawls and sloops, daysailers and deep-ocean cruisers.

"It just happened," Nat said. "When we arrived, in 1972, there was *Alabama, Shenandoah, Venture,* and us." "Us" being Nat and Pam, who'd left their former home on the island of Formentera in the Mediterranean, crossed to the Caribbean and spent a year sailing there with their toddler, Jessica, before heading up the East Coast. At the time, *Venture* and *Alabama* were rotting away on their moorings in Vineyard Haven Harbor, *Alabama* not even capable of a day sail. A few catboats remained (beamy boats with the mast at the bow, typically half as wide as they are long), as well as a few lingering Vineyard Haven 15s (21-foot sailboats that once raced here as a class). But that was it. The message of the era was, Nat said, "Get rid of that nasty wooden thing—go fiberglass, and all your problems will be solved." Many of the surviving wooden boats were sawn up and burned. Now, more than twenty-five years later, the harbor was packed with them, nearly ninety at the height of summer. In winter, all you saw was masts of varnished spruce reaching into the sky. Wooden boat owners are often the poorest yachtsmen on the water, tired wood vessels being the most affordable; thus many avoid the expense of hauling and storage every fall by leaving their boats in the water year-round, as Nat did. This is something you wouldn't want to do with a fiberglass boat, but it's fine, even good, for a traditional hull, since wood loves salt water—salt water *preserves* wood, and the hull doesn't suffer from drying out during the winter. The drawbacks are that the topsides and deck are subject to extreme weather, storms are more frequent, and floating patches of ice can mark the hull.

Often when he toured the harbor checking the boats, Nat would stop last, as he did this day, at his own boat. *Venture* had been given to him by one of the island's most respected sailors and wooden boat lovers, Pat West. The only stipulation was that Nat make her right again. Nat conceded that had he realized how much work would be involved in restoring *Venture,* he wouldn't have accepted, but he was glad now that he had. Pat West had sailed her when he was a young man in the 1920s and 1930s, when he would sway in the spreaders, spotting the swordfish then abundant in these waters for the boat's gin-swilling owner. When Pat bought her, *Venture* was more than twenty years old—considered an old boat—and by the time Nat got her, fifty years after *that,* she was a tired boat indeed. The planking was yellow pine from the southern United States, excellent material for a hull, but it had been fastened during its construction with iron nails. After thirty years or so, nails that sit in water rust, and the hull had become irreversibly punky and rotted with disintegrating iron all through it. Boatbuilders in 1910 were perfectly aware that iron, which was less expensive than bronze, rusted; evidently, though, they didn't believe that a boat would *need* to outlast the iron that held it together. And yet here she was, replanked and gleaming in the autumn sun, blue water sparkling, almost ninety years old, her current owner fixing a couple of cups of tea on the galley stove.

"That's another good thing about wooden boats," Nat told me, taking a seat on the port settee/berth beside the wood-burning stove, his mug steaming in the cool air. "People care enough to keep them. If this were a 'glass boat, no one would have cared enough to repair her. No one would have given her away. She wouldn't exist anymore."

Venture was a sweet boat to look at, and soothing to be aboard; sometimes Nat came here just to think. His younger daughter, Signe, twenty-five, had recently spent a summer living on her. The boat was a sixty-second row from the town dock, but once aboard you could feel absolutely solitary. If Nat was checking *Venture,* recharging her battery or performing any number of other chores on board, and if it was a fine day with a fair breeze, the strangest thing could happen. *Venture's* sails would go up when Nat's back was turned, and then, amazingly, she'd be off her mooring, and before Nat could do anything about it, she'd be clipping past the breakwater and heading into the sound.

Nat Benjamin had come to believe in wooden boats through experience. You could fix a wooden boat. A wooden boat could save your life in a storm. The new boats with their plastic hulls and plastic hardware weren't worth a damn at sea. He recalled one delivery, taking a brand-new fiberglass Bristol 40 from New England to Tortola. Midway he ran into a gale, not uncommon crossing the Gulf Stream in the fall. It blew for two days—not a terrible storm, but the boat took a beating. The foredeck hatch came loose because it was poorly built; water filled the bilge; the electrical and steering systems gave out; the batteries began to leak acid; bulkheads, the main 'thwartship structural pieces, popped out of the flexing hull. Nat eventually made landfall in Tortola, but when the owner saw his new purchase, he asked, "What did you do to my boat?" Young Nat replied, "*I* didn't do anything to your boat. The sea did it." She was all but totaled.

A grin formed now within Nat's shaggy, red-gray beard, his almond-shaped eyes squeezed, his teeth set, and he said, "You don't really understand wooden boats until you're in one and the wind lifts you up and slams you on your side and you charge through the waves."

October chop lapped against the side of the boat as Nat sipped tea, the unique sound of water against a wooden hull creating a peaceful backdrop for a man whose past was filled with adventure and yet who seemed to all who knew him to be deeply at peace himself. Wooden boats—the integrity of the things themselves and the work they required—were a sizable part of that peace.

"I feel pretty lucky to be able to make a living this way," Nat continued. "I wouldn't be a good house builder—all those straight lines." He'd worked as a carpenter during his first years on the island, since there was virtually no wooden boat work anywhere, and working at a fiberglass yard was unpleasant for him. He liked to use the work of the traditional carpenter as a contrasting example to explain his appreciation for wooden boats, to address in part the reason he had devoted the last twenty years of his life to building them. "One is straight, the other is curved," he said of the house and the boat. "With one, you order all the wood, it's premeasured and cut, and you just hammer it together. It's like a giant Erector set. With the other, every piece is handmade. Nothing is measured with a ruler and cut at a right angle. All the pieces are cut to size in relation to the other pieces. Nothing is measured the way we normally think of measuring. A house, you mark everything with a tape measure; a boat, you measure with battens and tick sticks. One is stuck in the mud; the other you can sail away on!"

✦

Nat arrived at work on foot each morning with his canvas briefcase, checked in with Ginny, returned some calls, and then headed
over to his largely solitary work at Mugwump. On Wednesdays he left
at one, aimed up Main Street toward home (his seaman's lurch recognizable from far away), ate some leftover beans or instant soup, and
spent the rest of the day at his drawing board. Nat was quiet as he
worked. A loud clearing of the throat and a "Let's see here" were his
most common sounds. Gentleness seemed to infuse his efforts, whether
he was bending in a frame or performing the finish carpentry of a
cabin trunk's dovetail joints: he gave the impression not just of working on a piece of metal or wood, but also of reflecting on it as he
worked it. And as he worked, fine evidence of his labor and mind
tugged at moorings in the harbor.

From the look of Nat at work in the shop or at the helm of *When
and If* or *Venture,* you'd never guess how unlikely this circumstance was.
And not simply that he had been designing and building plank-on-
frame boats successfully for nearly twenty years and supporting a family
by this work, beating hard but steadily windward, not only against
the currents of contemporary marine design and the marketplace but
against contemporary culture generally. Nat never studied naval architecture at a proper maritime academy. He didn't even graduate from
high school—he kept getting thrown out or disappearing on his own.
A boat bum for the first part of his adult life and more or less alcoholic
for most of it, an irresponsible young parent and husband by his own
solemn admission (off sailing when Pam gave birth to their first child
on a rural Mediterranean island), an entirely self-taught builder and designer, he was now one of the most respected men in the business; he'd
stopped drinking years before and was the model of the devoted family
man, as well as a loyal friend to many and active in community affairs,
especially when they concerned preserving this working waterfront.
Given another life, he said, he'd pursue something different, music or
writing, but in this life he was happy to build boats.

✦

Garrison, New York, was a leafy, provincial town when Nat grew up there in the 1950s, a commuter town where husbands in suits would take the train fifty miles south into New York City every day and return to the country in the evening. Activity swirled around the train station and the general store, the hubs of the town. There was only one black man in Garrison, and he collected garbage and shoveled coal—most of the furnaces were still coal-fired. There were plenty of kids to play with along the river or on the train tracks. When Nat talks about his childhood, he recalls the movie *Stand By Me*, based on a Stephen King story about a boy's coming of age in a small 1950s town. "That was my childhood to a tee. Walking on the railroad tracks, getting into trouble. Stephen King was born the same year I was, the same year Ross was, nineteen forty-seven."

While Nat himself terms his upbringing "middle class" (an assertion that one close friend and cousin calls "ludicrous"), the fact is that he is descended from political prominence and privilege more deeply blue-blooded than I could feasibly make up. His background is a history lesson in American politics and White Anglo-Saxon Protestant heritage, beginning with Peter Stuyvesant, the autocratic governor of seventeenth-century New Amsterdam, before it became New York—a direct ancestor whom Nat says he is not proud to claim. He is also a distant descendant of the Livingstons, a prominent family during colonial and postcolonial times. Robert Livingston sailed in a wooden boat to the shores of the British colonies of America in 1673 and then settled in Albany. Two of his offspring became members of the Continental Congress. The latter, Robert, was one of the committee of five men who drew up the Declaration of Independence.

A direct descendant of the Livingstons, Hamilton Fish likewise went into American politics, becoming, in the mid– to late nineteenth century, a member of the House, a governor of New York, a senator, and the U.S. secretary of state. His son, grandson, and great-grandson, all named Hamilton, were also distinguished members of Congress. And the next in line, young Ham Fish, Nat's cousin, is also a politician, having served as a state senator in New York and run for congressional office in two campaigns. Nat and Ham remain deep friends and often sail together.

Nat is also descended from the Bigelow family, whose most notable member was John Bigelow, a writer and diplomat, U.S. consul general in Paris during the Civil War, then minister to France, who, when he got done with all that, discovered and edited Ben Franklin's autobiography.

The Benjamins landed in America even earlier than the Livingstons, in 1632, on a boat called *Lyon*, emigrating from Sussex, England, and prospering almost immediately and for more than a century, until somebody bore a son with a talent for losing money. But this latter man, unfortunate though he was in the ways of finance, himself had a son named Park Benjamin, who made his own way in the world. At the age of twenty-six, Park abandoned ownership of a shoemaker's shop and became Captain Park Benjamin of the good ship *Prosperity*, a trading sloop that traveled between Norwich, Connecticut, and British Guiana, and this made all the difference for the family. Captain Benjamin became one of the most famous of the Norwich sea captains; the local paper was filled with stories of his adventures. Working just after the Revolution, a time of great maritime prosperity, many captains on the East Coast amassed fortunes by both exporting and importing. Park and his brother developed a plantation in Demerara, British Guiana. They arrived there with cattle, lumber, and flour from the States and left with sugar, coffee, and rum. Park bore a son in Demerara, also named Park, who caught a tropical disease that left one of his legs permanently shriveled. Young, lame Park would go on to have a career as a poet and editor and is chiefly remembered for bringing to wide attention a new author then in his early thirties, Nathaniel Hawthorne. When Park was thirteen, the family had returned permanently to Connecticut, but his father, Captain Benjamin, needing to oversee his holdings in British Guiana, had set sail once more from the East Coast in June 1824, bound for South America. He was never seen again. Pieces of the boat found by other vessels suggested that the ship had run into weather; young Park also lost his older brother in the wreck.

Nat's direct forebears seem to have been an effective group, by and large. But if it is true that we *become* what we do, and if what we *do* can possibly be transmitted by our genetic material in any way, then there is some logic in Nathaniel Benjamin's attraction to boats, and in the fact that the progenitor of this side of the Benjamin clan was a captain lost at sea.

Nat's paternal grandfather died in 1928 but left the family enough money to last through the Depression and late into the 1960s. Nat's father, William Hoffman Benjamin, known as Hoff, grew up in a "fairy-tale world," Nat says. Born in 1910, Hoff graduated from college smack in the middle of the Great Depression. "When the rest of the country was dying for work," Nat said, "he went off shooting caribou in Greenland with some of his well-heeled buddies from college. They got on a fishing boat in Norway and went to Spitsbergen and Iceland, shooting at everything that moved and having a blast. While the rest of the world was broke. He loved to talk about it, tell stories about that trip. That was his ultimate trip, that was his life as a young adventurer."

Hoff became a real estate broker in Manhattan. He and his wife, Joan, had their first child, Bunny, in 1941, and then later John, and last Nat. Nat thinks there may have been something unusual about the year 1947. Both his sister and his brother fell right into the world of their parents, while he rejected it. He was always hanging out with kids from the wrong side of the tracks, and by the time he was fifteen or so, he began to see his parents' world a little more clearly. John Cheever territory, he calls it. "I couldn't stand the people. My parents' generation. I got on with my parents OK—not that well with my mother—except I couldn't stand their whole generation. The whole double standard. I was brought up to call everyone Mr. this and Mrs. that as though they were somebody I should really respect. And then when you're fifteen or sixteen, all of a sudden you realize that he's sleeping with her, and she's doing this to him, and that's when I sort of said, 'I'm not gonna take it from you people anymore.' I just couldn't buy the whole act."

His parents sent him down to a prep school in Scarborough ("There were a few of us who wouldn't be tolerated in elementary school," he says of the local school system), but that didn't last long, either. Next he was sent to the Lenox School in Lenox, Massachusetts, another prep school. It wasn't all bad, he recalls. He loved being away from home, a sentiment likely shared by his mom (though they've since reconciled, Nat says, "I didn't get along with my mom from the day I was born"). He made friends easily and loved sports, but he couldn't abide the insufferable teachers. "I didn't perform well for them," he says, concluding, "Aside from the friends and the sports and the trouble I got into, I really didn't enjoy anything about school." His cousin Ham Fish says that "Nat was mischievous and uncooperative from an early age . . . a hellion."

So in 1963, at sixteen years old, he up and left, headed down to Texas, where he found work on a ranch in a small town south of Houston run by a well-known Texan, part American Indian, named Blackie Clark, a rancher and oil-field troubleshooter. Nat did tractor work and maintenance until he learned to ride, and then he held his own rounding up horses. At the end of the workday he'd sit in the shade of a pecan tree, drinking whiskey and playing dominoes with the other cowboys. He felt far from blackboards and homework, the hypocrisy of suburban New York, and he liked it that way. But Blackie took a genuine liking to Nat and, concerned about the boy's best interests, convinced him to return and finish high school.

Nat gave school another shot. His brother went to St. Mark's, a prestigious prep school, and then continued on to the University of Pennsylvania—which was what one was supposed to do, Nat had been told. The attitude in his family was, he says, "If you didn't go to a good prep school and an Ivy League college, you might as well go jump off a bridge." But was he going to school, he wondered, just so he could put on a suit and commute to the city for the rest of his life, like his father? That just didn't make sense to him, and so he left school again, never to return. He headed west, to San Francisco, where he found a job as a carpenter, his first extended work with wood. After a time he headed into the mountains to ski. Then he returned east and tended bar in Newport, Rhode Island, well before he'd reached legal drinking age.

From Newport, an East Coast yachting hub, Nat wangled a berth on a boat being delivered to the Virgin Islands, a 32-foot plastic sloop. But the captain was a drug addict who wigged out off the coast of Virginia. A whole new crew and a new captain flew down to Norfolk, and the previous crew were told to go home. And they all did. But Nat lingered on the dock. He had all his gear together and was ready to leave, but the new captain, Christopher Fay, looked him over. Fay sensed something interesting and unusual about this kid. He looked at Nat as he was about to go and said, "Why don't you stay?"

Nat did. He learned to sail by watching and by asking questions. Captain Fay took his time reaching the boat's destination, island-hopping in the Bahamas to the Abacos, across the Yellow Bank to Nassau, then the Exumas, and down to San Salvador, finally arriving in the Virgin Islands in January. By then the delivery captain had to return to the East Coast, but with the season just starting, the boat's owner wanted to charter it. He asked the captain if he knew of anyone. The captain shrugged and said, "How about that guy?" pointing at Nat. The owner asked if Nat was interested, and Nat said, "Sure!" He had been a cowboy, a carpenter, a ski bum, and a bartender, and now, still just twenty years old, he was the captain of a yacht.

Nat was having a blast.

As the captain of a boat, bopping around the West Indies, Nat was for the first time introduced to that renegade society of "wonderful and crazy boat people," as he calls them. The social upheaval of the 1960s was at full volume, and so there were many such renegades. Pam Stevens was one of them. A blue-eyed blonde from outside Boston, Pam had left art school to see a little bit of the world and had found work as an au pair for a Guadeloupean couple. She lived on a beautiful wooden powerboat that was tied up, by chance, exactly where Nat's boat was docked, in Yacht Haven Marina, St. Thomas. Nat had all but thrown the lines to Pam when he sailed in.

"Nat was very happy, always laughing," Pam recalls.

"Had nothing to do with the drugs I was taking," Nat adds.

"We were pretty wild and crazy," says Pam, averting her eyes.

"Vietnam was roaring along."

"The Beatles' *Sergeant Pepper's Lonely Hearts Club Band* had just come out," she says. "We had a record player on the boat, and we would pull it onto the dock and get a long extension cord and put it up on a table and listen to this record player." Thinking longer about Nat as a young man, a young captain, Pam adds, "Nat was brave."

"That's one way of putting it," Nat responds.

"I guess it wasn't just the alcohol, because even when he gave that up, he continued to be that way."

That year, when he was just out of his teens, subtle changes began to take place in Nat. Till then he had been both exuberantly reckless and smart enough to get away with it unharmed. But you can't live on a boat and spend your days sailing without learning something, and if you have an active mind and an observant nature, you can't help but begin to grow up. "I wanted to do a good *job*," Nat says, "and become respectable. Become respectable at something. I liked the *people*." And then he adds the critical piece of information about himself, the essential ingredient in any successful beginning, the trait that all success shares: "I liked the people who were *good* at it, who had good boats."

And so, blessed for whatever reason with the ability to recognize excellence, to watch it, to see how it was done, and to want to emulate it, Nat began his self-styled sailing education.

"Probably the most important lesson I learned was safety and sobriety at sea," Nat says. "Because I did have a bad grounding coming into Christiansted, St. Croix, at night. It was a bunch of crazy people on the boat, and everybody was stoned or drunk or both, and we had no compass light and the flashlights were not working, no engine, and sailing into this fairly complicated harbor. I'd never been in there before. I had a chart which I could barely read. We had to go through a narrow passage, then around this reef. It was something that now I just wouldn't *do*—at night without a good chart, without my compass light, I just wouldn't do it. But then, *anh,* no problem. And we hit the reef. We had gotten around one reef, and I had to make a sharp left turn and I didn't make it sharp enough, couldn't really see anything, and we grounded on this reef.

"It wasn't disastrous, because there wasn't a sea running, because we had already gotten around one reef. If I'd hit the outer reef, it could have been bad.

"We were fairly near town, and someone came out in a Boston Whaler and took my passengers, my friends, ashore and then they towed us off and we tied up at the dock. And the next morning I dove down to see the damage. There wasn't a lot of damage, sort of scoured the bottom of the keel, but it was a real eye-opener, a real awakening for me. I'll never forget it." Recalling it, Nat appears to shudder. "It sobered me up. From then on I never really drank much at all when I was sailing.

"If it's never happened to you, you don't know that hideous sound of grounding, grinding." His voice quiets. "It's a *terrifying* sound."

Nat made his first significant ocean voyage that winter, from the

Virgin Islands southeast to St. Bart's, an alluring French island they'd heard so much about from fellow sailors, and he and Pam weren't disappointed. "We go back there from time to time doing charters," Nat says now, "and it's a booming French tourist place. It's still beautiful, but it's jammed with people, the richest of the rich, and huge, fancy yachts, so I'm glad we saw it in nineteen sixty-eight. We were the only yacht in the harbor. There were other boats, but they were working boats, island traders, old power vessels."

He and Pam sailed back to the United States on an "English-built thing," Nat says, *Axel Heyst,* named for a character in Joseph Conrad's novel *Victory* who avoids all ties with the world. It was one of the first boats built using an experimental technique called cold-molding, a wooden boat held together by a lot of powerful adhesives. Nat and Pam returned to Garrison because Hollywood had descended there to film *Hello, Dolly!*—Gene Kelly directing Barbra Streisand—and there was easy money to be made. It was during that summer that Nat received what would be a fateful call from a friend in Garrison who knew a man who had a boat tied up in a marina in Malta, a 38-foot Block Island schooner named *Tappan Zee.* The owner wanted her delivered to Long Island. Nat called the man, struck up a conversation, and soon was offered the job of delivering it.

"I jumped at the opportunity," Nat said. "So in August, after the movie, we had a little bit of money saved up, we bought a couple of plane tickets, went over to Europe. We had some time, so we traveled around, bought an old motorcycle and went roaring through France, blew that up, and went hitchhiking the rest of the way."

Nat and Pam said good-bye in France, in the Alps. Pam stayed on to do some skiing with her friend Pamela Jane Street, and Nat hitched his way down to southern Italy.

Late one night he was walking through a remote Italian town with his big duffel, a sea bag, on his shoulder, when, with a terrific screech of brakes, a little Fiat came out of nowhere and plowed into him. Gas was expensive in Italy, and it was common for drivers to shut off their engines when coasting downhill. This driver didn't have his lights on, either. Nat never saw or heard the ghosting little car till the sound of brakes broke the night's hot stillness, by which time there was nothing he could do. He felt his body hit the hood and smash into the windshield, shattering it; he tumbled over the top of the car and onto the ground.

"*È morto,*" a voice above him said. Nat groaned, coming to. The crowd gathered around him realized that he was alive—his big sea duffel had likely saved him—and he was taken to a hospital that had no hot water, flies buzzing and crawling over everything, syringe-happy nuns dancing about, sick people moaning through the night. When he woke in the morning, he couldn't wait to flee. He found himself badly bruised, but nothing apparently was broken, so he lifted his duffel, hobbled aching to the front desk, said "*Arrivederci,*" and walked out. He was greeted by the man who'd run him down; he'd come to the hospital to find Nat. The man spoke no English, but Nat inferred from what few words he could pick up and from the man's tone—and also from the way he was jabbing a finger at his car—that he was demanding that Nat pay for the damage. The Fiat *was* pretty well totaled, Nat had to admit. As neither spoke the other's language, though, there was little for an unsympathetic Nat to do but limp to the side of the road and stick his thumb out. The first car that approached stopped to pick him up. The driver, wide-eyed, said, "I can't believe you're alive. I was there last night. I saw the whole thing." Nat, thus proven blessed, traveled south and boarded a boat for the short hop to Messina, Sicily. Then, Nat says, "I hitched a ferry over to Malta to meet my ship!"

V

Work on *Rebecca* continued through Thanksgiving, and the Christmas holidays came right on the heels of that. Nat built one beautiful teak deckhouse, Todd McGee built the other, and it was difficult to say whose joints were tighter. The deck framing was quickly completed, and work on the interior began shortly before Christmas. *Rebecca* was looking fine, as *WoodenBoat* readers could learn from David Stimson's feature article on her. Everyone was cruising, and the wooden boat world was watching. The new year came, 1999, and work slowed as people recovered from fatigue, overindulgence and excitement, and the flus and colds that ensued, but by mid-January work on the schooner had regained its stride.

Winter was the most productive time at the boatyard because there were few distractions posed by tourists, boat repair, or sailing. These were a boatbuilder's best months, January through April. Mugwump was wickedly cold, but work got done with the steady efficiency that comes from routine. A full year had passed since Nat and David cut out the first keel pieces, and now here was this planked-up, framed boat, hovering inside Mugwump every day. She was practically beginning to breathe, this *Rebecca*—and something more. People began to realize that she wasn't just another *example* of a great wooden boat; rather, she embodied and carried forth everything that wooden boats were all about—their strength and beauty, that combination of art and science that thrilled so many lovers of wooden boats. Her craftsmanship was exquisite. She would be an instrument, a 77,000-pound Stradivarius, just as meticulously built, equally refined, but also strong enough to take a beating at sea when called upon to do so.

David Stimson owned and played a violin made in the 1830s in the Netherlands (knowledge deduced from the shape of the F holes, the maple and tight-grained spruce construction, and the purfling, fashioned from baleen, a whale's plankton-catching fibers, David said, a popular Dutch decorating material of the time). This violin had been built to last, and violins were still, David would tell you, being built to last a century or more—maybe indefinitely, given proper care. What else was built to last that long anymore? Houses had been once, but no longer: owners of houses in Levittown were finding out that plywood delaminated and fell apart. It was a rare window today that had useful mullions; they were almost always flimsy decoration. Even churches weren't built to last these days. But *Rebecca* was.

✳

On February 8, 1999, a gloomy dawn emerged from a frigid night heavy with clouds and wind-whipped snow—the very heart of winter. The cold, dark Monday intensified the strange silence at Mugwump. Gusts whistled outside, and the smallest noise echoed inside this enormous shed. A lone man stood against a workbench, waiting for builders and work that wouldn't come. The man, a former cranberry farmer named Kerry Elkin who'd been forced to sell his farm in the aftermath of a divorce, had arrived today, at age forty-eight, to begin work as an apprentice boatbuilder. He was starting a new life in middle age and was looking for work that would be similarly meaningful to him—he needed to do something "real," he said, something durable and satisfying. But on this Monday morning when he had planned to begin that work, he stood alone in the Mugwump shed, arms folded against the cold, workboots deep in wood shavings still fresh from Friday's labor, and his warm presence in this shed heightened the sense of *Rebecca*'s abandonment.

All work had ceased as of Friday evening. Nat had called his crew and told them not to come to work on Monday. He explained that there was not enough money to meet next week's payroll, and none was forthcoming. The decision was made at the last minute, but the circumstance that led to the decision had existed even before construction began, and originated in the owner's inconsistent financing. Dan Adams had incorporated Mugwump Charters as the business that would fund the building of *Rebecca,* buy its materials and pay the boatwrights' salaries and pay Gannon & Benjamin, and now the money was gone.

David Stimson, who had moved his family from Maine to work on *Rebecca,* seemed the most hurt. Even worse than the financial loss—he would make significantly less money with which to support a family of four from the work he would do following *Rebecca*'s shutdown—were the disappointment and deflation of halting the creation of this magnificent vessel. "We all had a very good head of steam on it," he told me. "We all had our hearts in the project. Stopping was very difficult. It was hard to adjust."

Others took it in stride. Pat Cassidy, Todd McGee, and Casson Kennedy (a young boatbuilder with a second child on the way) moved to Five Corners Shipyard, where Bob Douglas had work for them. Todd even used the layoff to take a little surfing vacation in the Bahamas.

The one who deserved to be the most saddened, the most deflated, the most frustrated and angry, though, was the boat's creator. Nat should have wanted to throttle Dan Adams—foremost for disrupting his business and his livelihood and the livelihood of his crew, who might or might not return. These were his responsibilities, and he was not a man to take responsibility lightly. Nat couldn't call Dan to discuss the situation—Dan didn't even know that work on *Rebecca* had stopped—because Dan was in Tahiti with a new girlfriend and his buddy the actor Armand Assante, sailing. He was in the South Pacific *on vacation* while Nat was putting his crew out of work.

This kind of behavior infuriated Ross, who simply could not fathom how an adult could be so negligent about other people's lives. But Ross had been skeptical about Dan to begin with, a man who for the most part didn't do any real work—or none that Ross could see, at least, Dan's independent films notwithstanding.

Nat simply shrugged and said, "It'll get finished." Nat reacted to this new situation the same way he did to any other, with a mixture of philosophical resignation and practical intelligence. Yes, it was unfortunate, and it was not without financial repercussions for everyone, but as for the boat, *she would get finished.* She almost couldn't *not* get built. She was too far along, the evidence of her beauty and quality abundantly apparent. If Dan Adams couldn't come up with the money to finish her, somebody else surely would. There was not another construction like this one anywhere in the United States, or probably in the world. *Rebecca* would be built in her own good time. "A boat has its own life, its own destiny," Nat said.

Billy Mabie, a tall, dark sailor with the ribald humor of a man who's spent much of his life at sea, was to fashion *Rebecca*'s rig. He did hand-splicing and wire-rope work for G&B boats as a sideline to his main job as third mate on Exxon tankers traveling between San Francisco and Alaska. When Billy returned to the Vineyard after his most recent voyage, he wandered by the shop to say hello. When he heard the news about *Rebecca* and learned why, he issued a gravelly, unsurprised chuckle and spoke words that could well serve as the boatbuilder's maxim. Billy Mabie laughed, leaned against a workbench, and, in a loud, bawdy voice, said, "No cash, no splash."

That, of course, was what it all came down to, what so often trips up yacht construction. But yacht construction is usually initiated by a boatstruck man—a man, for instance, who would buy a boat without having a marine survey done—and a boatstruck man *never* has a clear view of his own finances, a realistic understanding of how much actual cash he has. The boatstruck man invariably has more cash in his head than he does in his pocket, so it's a wonder this kind of thing doesn't happen *all* the time.

Dan Adams, in addition to having a bad case of being boatstruck, had also engaged in two other money-devouring projects. His third movie, an independent film titled *The Mouse,* had not resulted in the kind of box-office receipts he'd been counting on, and he still had outstanding debts from it. The second undertaking was his restoration of a nineteenth-century brick barn up-island. Both projects were noble. The film had received respectable attention (including a favorable notice in the *New York Times*). He was saving one of the island's few remaining brick barns, and not by transforming the interior into some monstrosity of comfort for the idle rich; rather, he was keeping the interior true to the humble, rough structure itself. He might be careless about his finances, but he did seem to care about the right things—that is, old and true things, brick barns and plank-on-frame boats. Furthermore, he had been a perpetual presence at the boatyard, taking an active interest in the construction and in the lives of the crew, frequently treating everyone to lunch, showering them all with presents at Christmas. So while no one liked what had happened, all were forgiving. Dan had bitten off more than he could chew, and no one blamed him for that.

There was even something *appealing*—something daring and fundamentally unselfish—in Dan's boatstruck delusions and negligence: *Let's build it first and then figure out how to pay for it,* he seemed to be saying. *That way it's guaranteed to be built, which is the important thing, and I'll accept the consequences, whatever they may be.* And maybe Dan was right when he said a loan would be coming through any day.

In many ways, *Rebecca*'s shutdown underscored the inevitable conflict built into this work: the humble and honest craft of boatbuilding was practiced by relatively poor artisan builders for the benefit of wealthy customers. There was bound to be class friction in the world of wooden boats. While there was every reason for the crew to like Dan Adams and to be forgiving of his troubles, there were limits to their sympathy. He was a different breed of human from the boatwright. He had money given to him; he didn't know what real work was. He was the leisured rich. Work was forty hours a week for life; independent movies that didn't make money were a hobby.

The upper crust shared a similar dichotomy of emotion. Those who loved wooden yachts, and could afford them, embraced the boatwright, elevated him. To be in a boatyard was for them to come close to a life they often dreamed about. Many truly did look upon these boatbuilders and wish that in another life they might trade places with them, spend their years in a boatyard. More than a few gave regular and serious thought to chucking it all for the life of a wooden boat builder, for the chance to do something true and real and beautiful, to end this suffocating existence devoted solely to acquiring wealth, a life in which they touched nothing real all day long, didn't even trade in anything you could touch or hold in your hand, let alone sail away on.

The dreaming, of course, was typically done from within a fully loaded Lexus on the way down to Wall Street from the weekend country home on a Monday morning, or gazing out the window of the corporate jet on the way to a meeting in Atlanta. When they were honest with themselves, these wealthy men who could buy any kind of yacht they desired didn't *really* want to be that loser sanding boat hulls all day long, or stuck in a greasy bilge, lunching on cheap sardines

scooped out of a can with a scrap of wood off the shop floor, or freezing his ass off all winter long in some backwater Maine shack.

And the boatwright, for all his gratitude that there existed a
monied class to pay for this work he truly loved, was quick to denigrate
the fat cat, the crass materialist, who really didn't know anything about
boats and couldn't pick up a mooring to save his life, much less sail the
big boat he was having someone else build for him.

Nat, who in many ways straddled both worlds, had been around
boats long enough to know that the onerous rich rarely had a taste for
wooden boats. The ignorant nouveaus, the ones reeking of cologne and
B.O., jewelry flashing, the fat ones with hairy backs, they generally went
for the modern power yachts made of plastic. *"Stinkpot,"* Ross would
curse, scowling, when one crossed his bow on the sound. Stinkpots
were those gross, swollen tubs that plied the Vineyard Sound all summer
long, with their entertainment centers and loaded bars and ice machines
below, tearing up the water and trailing wakes of fumes and foam. In
other words, the human qualities necessary to acquire massive amounts
of dough were rarely compatible with owning and loving a wooden
boat. People who preferred wooden boats had a discriminating sensibility, a rarefied notion of human equality and equanimity toward their
fellow man that included an appreciation for the natural world and
natural things—sensibilities shared exactly by the boatwrights themselves. Wooden boats, Nat observed, tend to self-select their owners.
But even generous and forgiving Nat had a hard time defending that
claim when it came to *Rebecca*—he'd just chuckle and shrug.

"A boat has its own destiny," he'd repeat. "A boat has a life of its
own." And he'd lurch and swagger along Beach Road to work, canvas
bag in hand, rolled drawings protruding from it, and climb into another
boat, crawl down into the dark, greasy lazarette with his toolbox. Nat,
of course, with his long view of the situation, was right: the boat
probably would be built eventually, maybe even with Dan as the owner.

And she would be everything everyone had said of her. *Rebecca* was,
indeed, a structure of monumental significance in the world of wooden
boats—a big structure, one constructed using techniques hundreds of
years old, but at the same time a contemporary vessel built for pleasure-
cruising deep oceans. And she was being built to last a century and
more. As David Stimson had noted, this was the most important fact

about *Rebecca:* she was being built to outlast any other boat being built today, no matter the material. A couple of months weren't going to hurt anyone and would be insignificant in the whole life of *Rebecca,* no?

✳

Not long after David made his observation about *Rebecca's* durability, a Yale architecture professor named Vincent Scully addressed the very same issue in the *New York Times,* wondering what, if anything, that was built in the twentieth century would survive a thousand years, as some of the earth's great structures have done—the pyramids, for example, and Greek and Roman architecture, European cathedrals, the Great Wall of China. He proposed that prisons might be our legacy: "The confinement of wasted lives in immortal containers constitutes our most ambitious public work today," he wrote. He noted that the need to contain nuclear waste would also result in, it was to be hoped, permanent structures. And our highways—those, too, likely would last beyond the use of the machine that made them desirable in the first place.

Never before since the advent of cities themselves, Scully continued, had architects and builders embraced what he called an aesthetic of impermanence, choosing for their material "steel and glass, especially, along with light-frame structures clad with plastics and other things that blow away," compulsively designing structures "to outface Nature and to encourage perpetual revolution, in which nothing could be decently allowed to last too long."

Our current generations, Scully allowed, might shrug at such a notion and not think it all that bad; might say, "So what, so what if something's built to last only thirty years?" This would not necessarily be a cynical reaction, but rather the slouching acceptance of fact: things fall apart. This is the linchpin of a modern sensibility. They might go so far as to call it presumptuous and pretentious to desire to build something that could last a thousand years—*You think you're that good, your work is that important, that it will be valued by civilizations a thousand years from now?*

But there were others who found something terribly sad about this—that an entire culture of combined races and generations, a mass of people, *the world,* increasingly dominated by American culture, could embrace an aesthetic of impermanence. Wasn't it tantamount to saying your work didn't matter, your work and therefore your life, that they were disposable, that you could throw them away yourself or let time cut them down almost as quickly?

"Perhaps for things to last," Scully suggested, "one has to love their physicality in old, premodern ways. But can we ever love an individual building again as the people of the eleventh century loved their churches?" He didn't answer that question, but he underscored the statement before it by concluding, "Eros builds everything that counts and takes care of it."

So it was with builders, those builders along every coast from Maine to Washington who built boats out of wood. Why else would they be doing it if they weren't guided by Eros, working for the sheer love of it? A tiny few, too few to count, continued to build wooden boats at the height of America's all-but-absolute embrace of fiberglass, beating hard into the wind of contemporary culture, because they loved wood and loved boats built out of it. Such boats were gratifying to make; you could run your hand over your progress at the end of the workday. They were satisfying, solid shells on water. And no one who loved them could avoid extending their meaning beyond the physical— just the *idea* of a wooden boat was enough to excite the blood. The first log rafts were built tens of thousands of years ago, and the first planked boats extended back to the Bronze Age. These creatures sometimes did seem literally to breathe; they were connected to our origins and could connect us back again, if we let them. They were built with material of this earth, often using ancient tools (adzes, levers, and rollers). The fundamental intelligence and integrity of the concept allowed writers such as *WoodenBoat's* Jon Wilson and Maynard Bray to propel these boats genuinely into the realm of symbol. They represented solid traditionalism, not fleeting, counterfeit modernism, these writers said. Wooden boats moved contrary to our high-speed, synthetic culture. They were a symbol of and a metaphor for excellence. They combined in their construction art, science, and—given their material, the medium they sailed on, and their imitation of aquatic shapes—the natural world. Wooden boats, small daysailers and seafaring yachts alike, were built by hand and by eye, and in this they represented mankind's enduring ability to create complex, practical, and beautiful structures on an elemental level.

So it was far more than that they were heavy and safe at sea, comfortable, infinitely repairable, and capable of achieving a greater physical beauty than anything else that rode on water. People who had been around wooden boats for any length of time began to speak of the "truth" of wooden boats and to claim there was wisdom in them. Wooden boats, it was said, taught you about yourself. The "truth" of a wooden boat, Wilson wrote, "is the integrity we sense beneath the beauty."

It all came down, in the end, to those curves, those lines that boatstruck you, those tantalizing, never-ending, just-beyond-the-horizon curves, in the sheer, in the sails, everywhere on the hull, everywhere on the boat, those curves.

The novelist Mark Helprin, describing the majestic bridges connecting Manhattan to Brooklyn and to New Jersey in his novel *Winter's Tale,* wrote, "They extend and embody our finest efforts," and made special note of the catenary, the sloping bar of a suspension bridge: "The catenary," he called out with the volume of Walt Whitman, "this marvelous graceful thing, this joy of physics, this perfect balance between rebellion and obedience, is God's own signature on earth."

But Helprin got it slightly wrong. Awestruck by the architectural magnitude of the bridge, he missed the important distinction: it is not the catenary of the bridge, that massive work of steel, but rather the *curve* generally that is God's signature. And part of the greatness of that signature is that it can be of any size, large or small; it can show up in a skater's approach to a double axel, etched in ice, or in the trajectory of a center fielder's throw to home plate, or in the arch of your daughter's instep. The curve is the signature. And the curve of a catenary is ultimately not as intriguing and complex as the curve of a boat because, as Ross Gannon was quick to point out, a catenary is always the same shape, dependent on and determined by the height of and distances between the bridge towers. Indeterminate and dependent on the eye of the builder, the curves of a boat define its art and its grace. These curves, this signature, are everywhere on a wooden boat, they are what make men, and not a few women, crazy when they get near a true and beautiful boat.

The day up in Rockport when I first met Nat and Ross, I'd asked Jon Wilson what it was about boats that so captivated men, why the magazine he'd started was successful beyond all proportion, what made boatstruck people do ridiculous things and lose their judgment. Jon had paused thoughtfully and said, "I think wooden boats touch men's dream life."

That was it, of course—there had to be a reason for the lure of these vessels, a lure crazily disproportionate to their actual worth and practicality. They were more than just a boat; they transported you across more than just water. They could connect you not simply to a different life, or a better life, but to your dream life, the greatest end, coherent bliss, with everything unknown answered. This was it, this was why men actually became slightly mad when they were around boats. It was then that they were as close as they would ever be to God's signature—the wooden boat—then that they could sense a path to their dream life, to heaven itself: they could sail away on it. Whatever they knew heaven to be, they knew just as certainly that they could get there on a beautiful wooden boat.

Rebecca embodied all of this: the art, the science, the capacity for metaphor, the connection to the natural world, man's ability to dream. And she was being built—as almost nothing on earth was anymore except perhaps her opposite, the American prison—to outlast the generations that were building her, to outlast even their grandchildren. She was, to the builders, to the people who came to visit her, to the people who watched her coming together, and to the people imagining her imminent launch, nothing less than the last kind of cathedral our culture had left, a heaven-bound cathedral of curves.

✳

On February 8, 1999, work on *Rebecca* ceased. There was no more money; vendors hadn't been paid, no paychecks could be written, and, with the owner vacationing, no one could say when any money would be available. Money was the operative force here, and that force was gone.

Talk all you want about boat as metaphor, as heritage, as a magical combination of art and science, as a floating *cathedral*. As *God's own signature*. This was America at the end of the twentieth century, and America at the end of the twentieth century had something called a bottom line, and the *bottom line* was this: *No cash, no splash!*

Two

Elisa Lee

I

Ross Gannon is searching for a knee. After nearly an hour in a patch of mud and brush nicknamed Bargain Acres, he has found the husky trunk of a black locust tree, maybe fifty years old, close to 5 feet long and 1½ feet in diameter before it widens into its roots. Ross thinks the knee is inside. It's a good piece of wood, though the heart rot in locust always concerns him. You never know till you cut it open. Ross wrestles the monster log into the back of his beat-to-hell pickup, a small, low-riding, wood-sided, diesel-engined buggy painted black, with "Pepe's" on the door and crude red flames on the hood. He throws the truck into gear and crawls over the crests and troughs of the rutted road out of Bargain Acres.

Ross crunches to a halt on the gravel of the G&B dirt lot and grabs a chain saw from inside. He wants to take some weight off this hefty trunk before he tries to haul it up the stairs—enough excess wood so that it will fit through the ship's saw. He pulls the start cord and is soon obscured within a cloud of smoke and flying wood chips, the noise drowning out all other sound, the smell of gasoline and fresh-cut wood mixing with the briny air rolling in off the beach.

Although reduced in weight and shape, the stump still requires two people to lug it up the steps and into the G&B shop. Duane Case helps Ross heft it onto the ship's saw. Ross explains to Duane that the center will be rotten straight through—inevitable with locust, a member of the pea family, a legume—so they can cut the stump in half, reducing its weight considerably, and work it from there. It's early February, the day is dry but cold, and what heat there is in the shop emanates from a small wood-burning stove beside Ross. He reaches up to the spoke of the upper wheel of the ship's saw, then presses a black button on a box mounted on the wall as he pulls down on the spoke to get the heavy machinery going in the cold. The motor groans, the wheels stagger and then slowly hum, and in a moment they're spinning and the inch-wide blade whirs and screeches. Supporting the big log at its top end with his arms and pushing with his chest, Ross directs the root end of the stump through the blade, lowering the stump when the end passes through the blade guide. Duane, on the other side of the saw, guides and heaves on the stump until the blade is through. Then the two men lean hard into the stump to push it through the blade again.

"Oh, boy," Ross says, staring at the second cut. "More hollow stuff."

Many knees will go into the new boat to connect pieces per-pendicular to one another, but this knee is an especially critical one: the transom knee connects the transom to the keel at a virtual right angle, on an object unfriendly to right angles. Its eventual shape will be similar to that of a giant wooden shelf bracket—triangular, almost L-shaped. The short end will be fastened to the keel, the long end to the transom. The key to the success of this piece of wood, what makes it so particularly valuable to the boat, is not its hardness (though locust is sometimes so hard that screws break off in it) but its grain. Ross has spent half his day searching for a log that will contain this piece, and now he is painstakingly sawing off excess around it, like a sculptor tak-ing the first chunks off a block of marble, so that the knee will have a grain that starts out vertical and bends 90 degrees until it's horizontal. A curve built into the wood itself, where the trunk descends into the roots, will result in a strong piece fixing transom to keel.

Ross and Duane slide the stump, diminished but still requiring two people to work it, back around in front of the blade, turning it on its side. They stare at it, and Ross says, "Let's go across here," motioning with a knotty, bent index finger marked with permanent black creases. He's trying to work it down to size without removing anything they might need, or ruining the good stump.

Soon the stump of black locust is reduced to a fat triangle—part trunk, part root, more than a yard high, half that long, and about five inches thick. Duane and Ross stare at the piece of wood; Ross holds the plywood pattern of the knee on top of it. He uses his collapsible ruler to check its width. The piece is still mottled with rot and wormy sapwood covering. Ross says, "I don't know if we can plane it to three inches, but let's see."

Ross hefts it to the planer behind him. He cranks the lever to the thickness he wants, hits the button, and sends the wood through. The volume and vibration overpower the senses—together they dribble your brain inside your skull like a basketball. Ginny, in the office at the opposite end of the shop with the door closed, wears earphones when the planer is in use for any length of time. It's one of the most frequently used pieces of equipment here. This knee is short, so that buzz-saw rattle, which could serve as the town fire alarm, lasts only about twenty seconds.

"It's a little sharper than it was yesterday," Duane says of the blade.

"And this isn't angelique," Ross adds. Angelique is nearly as hard as the blades themselves.

Ross flips the wood and sends it through again. And again and again, reducing its width by ¼ inch with each pass. The wood looks better and better, and they have uncovered no unforeseen knots, cracks, or rot. Ross places the quarter-inch plywood pattern on top. It seems to fit, just, and he traces the pattern on the wood with his pencil. "That's really not too bad," Ross says. "Why don't we cut it out and see?" They return the piece to the ship's saw and cut it along the pencil outline. All surfaces appear clean, with only a little bit of sapwood on the inner curve, which can be removed with a spoke shave.

Duane, who's been working with Ross only because he happened to be passing the ship's saw when Ross needed help, returns to his solitary work—finishing yet another Bella, the fourth one made to Nat's design for a 21-foot sloop—leaving Ross alone with his knee. Ross is pleased with his morning's work. It was entirely possible that as he planed the piece of wood, he'd uncover a black divot that would grow with each pass through the planer and render the piece unacceptable—meaning he'd have spent a whole morning seeking, lugging, sawing, and planing firewood. Instead he has a good piece for the new boat, a piece taken out of a log that would have been left to rot had Ross not scavenged it. Trees are taken down all over the island to make room for houses, and Ross collects as many of the big stumps as he can of the good wood—locust, white oak—and throws them in the brush until he needs them. Sometimes they don't amount to anything, but this tree would now likely live for decades inside a boat.

Ross hands the knee to me—I've begun by this time to show up for work—and tells me to finish it with block plane and spoke shave: the two flat edges which will attach to the transom and keel must be perfectly flat, and the sapwood in the curve must be removed and its edges rounded off. It is among the first pieces of the new boat being set up in the outer shop of Gannon & Benjamin, the yard's first power-boat, commissioned by Jonathan Edwards. Currently the outer shop is just a big white floor with a 32-foot picture of a boat on it. Ross finished this lofting on Monday—the day *Rebecca* was shut down—and the first pieces of the new boat's backbone are already beginning to appear.

This boat, too, would not be under way were it not for Jonathan Edwards's friend Dan Adams. The summer before, Jonathan, a tall, lanky singer-songwriter with clipped hair, balding on top, and a short mustache and beard, decided to buy a boat. He had legions of fans in the acoustic-guitar, folk-oriented segment of the music industry, but he was still known popularly and widely for his early 1970s hits "Sunshine" ("Sunshine go away today / I don't feel much like dancing") and "Shanty" ("I'm gonna lay around the shanty, mama / and put a good buzz on"). He had written and recorded music for Dan's last movie, and he continued to perform at small venues throughout the United States. He was fifty-two, divorced, a granddad, and when he wasn't touring, he lived on St. Croix, in the U.S. Virgin Islands. Jonathan wanted a simple little island-hopping vessel, a little bass boat, he thought, and he'd been up in Maine looking for one. Jonathan had played the Vineyard almost yearly since the 1960s. He'd usually wander through the boatyard when he was in town. He was drawn to it, as so many were; it was a part of the island. Nat and Ross knew the musician but never made a fuss over him, which was their way and the boatyard's way. Ted Okie would howl remembering how terse, nearly to the point of rudeness, Ginny had been with the pop music entertainer Billy Joel when he came by once asking for information, but Joel was fortunate to be treated like any other; it would only have enhanced Ginny's skepticism to know he was somebody famous. When Jonathan visited Dan on the Vineyard in the summer of 1998, Dan spurred him on—with the sight of the schooner project under way, all that incredible wood, his enthusiasm. If Jonathan intended to buy a boat, he had to at least talk to Nat and Ross.

"Dan really talked me into getting a wooden boat," Jonathan says now. "I looked at the job they were doing on the *Rebecca* and I decided that, ya know, a wooden boat would really be the best of all worlds." Jonathan notes that Nat and Ross's shop, who they were, the way they worked, had always been seductive to him. "They're a phenomenal team," he says.

The three men had lunch outside at Mystic Grill, a lunch joint on Beach Road between Mugwump and G&B, and Jonathan described to Nat and Ross the kind of boat he was looking for, how he intended to use it, and what waters it would cruise—a boat to hop around to other Virgin Islands near St. Croix in, a boat strong enough to comfortably motor across to Florida and perhaps run up the Intercoastal, to cruise as far north as Maine. Then he drew a picture on a napkin of the kind of boat he wanted.

Nat went home, sat down at his drawing table, and made freehand sketches of four boats. The first three were all bass boats, stout fishing boats with small deckhouses, in the style and length Jonathan had requested. Nat would have designed any of these three boats if that had been what Jonathan really wanted. But Nat didn't think Jonathan really *needed* a bass boat, and he didn't really *want* to build any of the first three boats he drew. They were too small to take out on the ocean, one of Jonathan's criteria. And so Nat sketched a fourth boat, a boat that he sensed from the conversation would better suit his prospective customer's stated purposes, and a boat that he himself would much prefer to design and build—a 1930s-style workboat, not dissimilar to a lobsterboat, but with a raised deck that would provide a dry bow, an important factor in the choppy waters around the Virgin Islands, and plenty of room below, since Jonathan was so tall.

Lobsterboats began as big rowboats, dories and peapods, when lobsters were so plentiful on the East Coast that they could be scooped up by hand from tidal pools or gathered out of shoal waters with a gaff. In the early nineteenth century an ingenious invention called a wetwell—a box built into the hull of a boat that allowed ocean water to

circulate freely—enabled lobsters to be transported from Maine to
Boston and New York, thereby turning lobstering into an industry.
Sailboats called smacks, small gaff sloops, were built for the work. The
gasoline engine, which began to appear at the turn of the century,
transformed the boats' design: sloops lost their centerboard displacement hulls and were flattened out in back to support the weight of the
engine. Just after World War I, Will Frost enlarged a skiff into what
would become the classic Down East lobsterboat, and earned himself a
name as the grandfather of that style. Sloops were by then obsolete.

The boat Nat drew looked not dissimilar to the classic Down East
lobsterboat, but his boat would be beamy, its width measuring more
than a third its length, and it would have a different underwater profile
from the Maine version, resembling instead a Novi, or Nova Scotia
lobsterboat, to which Nat felt it was more closely related. Nat didn't
want to pigeonhole it in any case; he preferred to think of his design
simply as a classic workboat of the 1930s era.

"I'd seen dozens of them in magazines, and occasionally one of
them shows up in the flesh," he said. "I liked them because they're
practical. They give you space. The raised deck is cheaper to build than
a separate deckhouse, so it's more cost-effective. It has a neat look to it.
And it gives you that big, dry bow, too."

The men met next in the plain, small Mugwump office. Nat
showed Jonathan the first three sketches—small, a little frumpy—and
then, after appropriate discussion, reflection, and a few tepid, courte-
ous remarks from Jonathan, he revealed the sketch of his powerful retro
workboat yacht, and the musician began to blink. This was a fine-
looking boat. He liked it. He liked it a lot. Then Jonathan mentioned
something he'd been thinking about for a while. This might sound
silly, and they might show me the door, he thought to himself, but
then he wondered aloud, tentatively, Was there any chance of having some kind of rig, even a small one, that would let him sail and
conserve fuel, or that he could use in the event that the engine
conked out?

Nat and Ross, to Jonathan's surprise, said, "Great idea!" Ross
pulled the pencil from behind his ear and drew a mast with a gaff rig
amidships, right there on Nat's drawing.

"We started talking this thing up," Nat says. "We told him why this would be a better boat. And Jonathan got more and more excited." To anyone who knows Nat very well, it would come as no surprise that Jonathan Edwards would convince Nat that this was the best boat possible for him.

Because Jonathan had a tight budget, the boat was limited to 32 feet. Nat wanted to design and build this boat, and so, slightly underbidding to accommodate Jonathan's budget, he and Ross agreed on a price of $85,000, with a bare-bones interior and no rig (the rig could be made anytime after the boat had been completed). The deal was struck. Construction would begin the following winter, aiming for a summer launch.

And so in this first week of February, Ross lofts this boat, and next week he will lift the first plywood patterns for her backbone.

✷

Even in the midst of a routine day, Ross has an otherworldly look about him, due in part to his bushy beard (moving from brown to gray), his cool blue eyes, his chiseled nose and tight, angular teeth. Although there seems to be not a soul on the island who does not admire and respect Ross, even friends use the word *complicated* when they try to describe him. He is invariably defined in contrast to or in comparison with Nat, the dashing sailor, boat designer, and family man. "Nat is a gentler person, a kinder person," says one friend who is close to both. "His family life is stable; he adores his daughters. Ross is, uh—it's a bumpy road for Ross, and it always will be. He's hard. He's difficult. . . . He's very judgmental, very critical, very hard on people. He's very hard on himself. He works *so hard*."

Another friend who has known Ross for twenty years, and Nat not as long, says, "Ross is a tougher read. Ross tends to dwell inside himself more."

His physical attributes and the "complications" of his personality combine to give you the impression that Ross is not completely of this time. Spend long enough in the yard and it will at some point dawn on you that his face, his whole physical presence, are straight out of the Civil War era, with those pale eyes and that bushy beard. Against the backdrop of the ancient machinery of the shop, with a knife in his hand, or an adze—not an uncommon sight—he is a Matthew

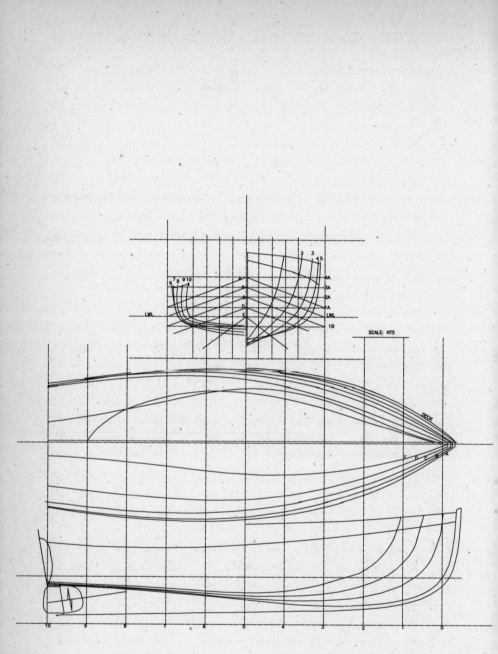

SCALE: NTS

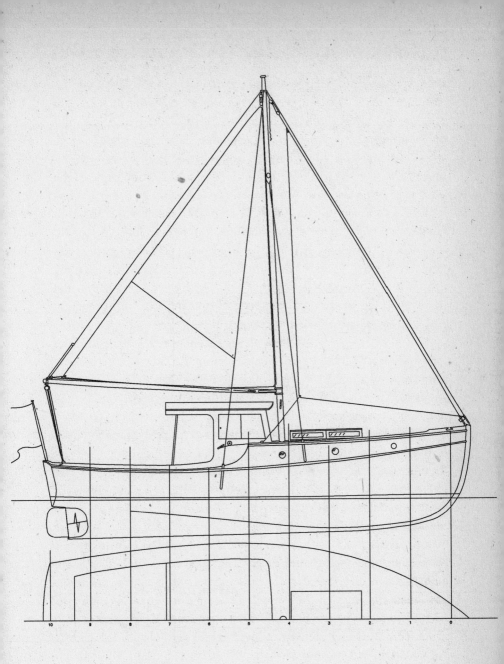

Elisa Lee lines drawing, *left,* and sail plan, *above,* by Nat Benjamin

Brady photograph, deep sepia and crumbling at the edges. Even his clothes wouldn't date him, they're so plain—a long-johns shirt beneath a heavy wool vest, both worn smooth from work. Ross stands just under six feet; he is strong and fit, cheerful, unfailingly friendly to all who enter the shop, and willing to stop his own work to show someone a faster way to cut a dovetail joint by using the tilting blade of the ship's saw, or to diagnose a boat owner's concern about a leaky sternpost. He's happy in his work, feels lucky to have fallen into it and even a little amazed that he, restless by nature, continues to love it so much.

"The thing about building boats of wood," Ross says, "is you never really get as good as you want to be at it. It's forever challenging. Furniture makers approach perfection. Their joints get tighter and tighter, and the pieces are more perfect. And boats are the same way. The more you do it, the better you get."

He grins, his mouth full of sharp, bright teeth. "But you don't take a piece of furniture and thrash it around in salt water and sunshine. If a furniture maker took his dining room table and went out and rowed it around in the harbor, and then let it sit out in the sunshine for a couple years, what would be left?

"The challenge of making something that can withstand the strains of the environment, I find challenging and forever interesting."

For Ross, the strength in wooden boats is not what he truly loves. He's devoted to wood and despises fiberglass—the material itself, boats made out of it, and all that it represents—but he will also be the first to admit that you can't make a blanket argument that wood is stronger than fiberglass. All boats move and flex; a wooden hull is composed of pieces and therefore has a lot of give, but a monocoque hull, a single molded piece, can potentially be built as strong as if not stronger than a wooden one (though this is rarely the case in practice, according to Ross, who feels that the fiberglass boats being built today are not only structurally weak but chintzy besides). So strength is a critical part of the equation, but it's not what draws him to the work. What draws him to the work is beauty; and what keeps his mind and hands continuously and intensely engaged is strength's relationship to beauty.

"The most significant part about a traditionally built wooden boat is that it's pretty," he says. "Even a good metal boat with wooden trim doesn't stack up. Fiberglass always looks like it came out of a mold. Cold-molded does, to me. But when you see the members of a wooden boat, what's inside of it, outside of it, you can see the stem, you can see the various pieces—it's just a prettier thing to behold than any other kind of boat, and that's the root of it for me.

"I *love* the intermingling of function and form," he continues. "It *has* to be beautiful, it has to look delicate, graceful. And yet it's got to do this *job*. An important job. You don't want to jeopardize anybody's life. You don't really balance those two"—beauty and function—because "one carries more weight. But you're trying to bring the other into view."

Ross is glad at last to have started on the new boat, referred to in the yard as the lobsterboat or the powerboat, and you can see it in the buoyancy with which he approaches each new task. For the past year he's been the one stuck in the shop, dealing with the minutiae of the business, with boat repair and small construction (a 7-foot skiff, *Little Love,* is almost complete; he didn't even need any drawings for that, just lofted it out of his head and onto the floor), while his partner has all year long been working on the sexy schooner. Ross concedes he was a little jealous. But last week, the first week of February, he got down on his hands and knees with the drawings, a table of offsets, a ruler, and a pencil, and he laid down the lines of the new boat. For everyone, it's an exciting time—far more thrilling and fraught with anticipation than any other repair or construction since the lofting of *Rebecca*. It's almost like finding out that the boatshop is pregnant—a new boat will soon enter the world.

II

The young Frenchman, Jim Bresson, has been given the stem, the foremost edge of the boat, to construct. Although serious and focused on the wood and tools at hand, Jim is the most jovial spirit in the shop. His short, dark hair stands on end first thing in the morning, as if his mattress were somewhere in the shop. He scratches his scalp, clamping one eye and smiling groggily, a dustpan in his hand. "Heh-looow," he says as his colleagues arrive one by one. He's tall and slender (not one of these boatbuilder-sailors carries excess weight) and is among the first to turn up in the morning. Jim wears a brown winter vest over a hooded blue sweatshirt over a green turtleneck, with brown work pants and brown laceless workboots. He sweeps the shavings off the benches as Duane stokes wood shavings to start the fire; then he sweeps the floor, organizes the tools that were left strewn around the shop the night before, and gets to work.

Six years ago, about to graduate from a wooden boat school in Marseilles, Jim asked a friend to keep a lookout for wooden boat yards as he traveled around the East Coast of the United States. This friend happened to spend a few days on the Vineyard, where he wandered into G&B, met Nat, and told him about Jim. A few overseas faxes were exchanged, and Jim appeared at G&B not long after. A year later the death of his father forced him to return to France, where he bummed around until his money ran out, then developed a wooden boat project with a friend—the building of a racing boat designed in the late 1800s by the French Impressionist Gustave Caillebotte, who some claim revolutionized French racing yacht design. Not long after that project ended, Jim returned to Martha's Vineyard, now age twenty-seven, eager for more work at Gannon & Benjamin.

When Jim isn't thinking about boats, he's thinking about food. Apparently deep in concentrated work, he will stop, lift his head, and announce to anyone present, "I think this weekend I will make a couscous. With lamb sausage." He will describe the kind of sausage he would buy were he in Paris. Then he will grin at the thought, and then laugh. He is fond of rum, and he is fond of women (whom he is inclined to date in an overlapping manner, with the inevitable consequences). He is seemingly never without a smile, and laughs and jokes continually through the workday. Soon he will begin arriving two hours early at the boatyard to get in some extra work on his own boat, which he bought with the help of Brad Ives: a 25-foot Folkboat built in Hong Kong called *Tomahawk,* the slender, teak-hulled bullet sits on jackstands beside the shop, a tarp hiding the extraordinary amount of work that must be done before the vessel will even float, let alone carry Jim to the coast of France, his ultimate goal. At night he reads a book Ginny has loaned him called *The Atlantic Crossing Guide.*

The three pieces that will form the new boat's bow—a stem and forekeel connected by a long triangular knee—are all clearly drawn on the lofting floor. Jim slides to his knees with a hammer, nails, jogged fingers, and some large pieces of quarter-inch plywood and begins pounding in the jogged fingers pointing to the line of the stem. He slides plywood beneath them, positions a batten, and marks the first edge of the piece. When he has the piece's complete outline on plywood, he cuts it out on the band saw. It's nearly five feet long and has just a slight bend, since Nat has designed this boat with a nearly plumb stem.

Two parallel lines run nearly the entire length of the drawing of the keel. The most important lines on the lofting floor, they represent the rabbet, the V-shaped groove into which the planks will fit, joining the planks to the backbone. All other lines are relative to them. Jim must now mark the rabbet on his pattern, which he again does by using the jogged fingers, positioning their ends above the rabbet line and sliding the plywood pattern underneath to re-create the lines.

As Jim finishes the pattern, Ross hauls a big block of angelique into the dark shop, 4 or 5 inches thick, a foot wide, and more than

5 feet long, a timber that Ross thinks will make a nice stem. He maneuvers it onto sawhorses and stares at the wood. He stares some more.
He borrows Jim's pattern and lays it on the wood. He stares, adjusts the
pattern, takes the pattern off, stares, returns the pattern, and eventually
retrieves the pencil from behind his ear and, his long, knotty, black-
creased fingers spread out on the plywood to hold it still, outlines the
bow-stem pattern, having, he thinks, made optimal use of the wood
and its grain for the gentle sweep of this stem.

The two men push the pencil line on the angelique through the
whirring blade of the ship's saw. Jim hefts the pieces to the bench, se-
cures the stem in a vise, and begins to take strong sweeps with his plane
across the surface that will be bolted to the knee—it must be perfectly
smooth, must fit like onionskin. Jim checks his progress with squinting
eyes and a carpenter's square.

When he has finished planing the stem's flat surfaces—the places
where it will connect to the knee and forekeel—he draws and then
cuts out a notch in the center of the surface using a Skilsaw and a
chisel. He will cut an identical notch in the knee, and when the two
parts are fastened together, a piece of wood, called a key, will be slid
like a deadbolt into the hole made by the two notches, all surfaces ending up perfectly flush. The key, a simple joint, will help to prevent any
shifting between the stem and the knee.

When the key has been cut, Jim can begin work on the critical
groove called the rabbet, ⅞ inch deep, with an undulating bevel that
will run the length of the keel.

✳

Bob Osleeb is outside the shop, building a frame for the transom.
Bob, too, uses a full-sized pattern taken off the lofting floor, and he
will do the first bending of wood. The back of the boat will comprise
two layers of 8-foot planks of ½-inch-thick wana. It will have some
rake—tilt slightly back—and will curve out about 8 inches. Bob builds
what looks like a small bridge out of 2-by-8's and 2-by-4's; he will
bend wana planks over the hump of a bridge that he's built, clamp
them down, and glue them edge-to-edge and back-to-back. This is
called a laminated transom, the same kind used for *Rebecca,* and it's as

close as this yard will get to that epoxy-and-wood hybrid referred to as cold-molding.

Bob is, in fact, a fan and happy advocate of cold-molding, making him a rare soul in this traditional yard whose articulate proprietors dismiss the method with confident laughter; to them the claim that a cold-molded boat is a wooden boat is as valid as a counterfeit sawbuck. Bob built his last boat cold-molded, a 24-foot open boat. He believes that the method of overlapping strips of wood glued together can make for a strong boat, and furthermore, it requires no heavy machinery or enormous timbers.

"I'm an open-boat kinda guy," says Bob, forty-five years old and a compact five-nine, with a graying beard on his round face and a ball cap covering his wavy brown hair. In open boats Bob, his wife, Marilyn, and often their youngest son, Dylan, now twelve, sailed the waters of Lake Huron and Lake Nipissing, and made the passage from Georgian Bay in the middle of Canada to Martha's Vineyard, via Lake Ontario, the Erie Canal, and the Hudson River. Their most interesting sailing was in the Strait of Georgia, off British Columbia, when they lived out there. Their worst sailing happened on a trip north from Florida in an 18-footer Bob also built, when they were hammered continuously by heavy winds for two and a half months, their mast breaking three times before Bob finally gave up and hauled out in South Carolina.

It was in 1990 that Bob read an article about Gannon & Benjamin in *WoodenBoat* and decided to sail over to have a look. G&B offered him a job, and he accepted. After a six-year absence from the yard, having heard about the schooner and the powerboat, Bob returned to the Vineyard with Marilyn last fall.

"I think they're angels," Bob says of Nat and Ross, without a smile or any hint of self-consciousness. "Wooden boat angels."

Bob takes less than a week to complete the transom, which is then set outside the shop on sawhorses near the friction winch—a big, curved construction of light-brown wana and glue, 8 feet wide, 4 feet high, and an inch thick.

✳

While Jim carefully carves the rabbet with chisel and mallet, and while Bob fashions the transom, Bruce Davies and Ted Okie make patterns for the double-sawn frames, the ribs of the hull.

Nat and Ross conceivably could have specified only steam-bent frames, in which case the powerboat would have been made the way sailboats of the same length are constructed. Interior molds of inexpensive pine would be set up along the keel at each of the ten stations, mimicking interior frames; spruce ribbands would then be bent around the molds where the planking would eventually go, and thick strips of white oak would be cooked and then clamped against the ribbands to form the framework of the hull. But the design calls for frames sawn out of two layers of angelique or locust on each of the nine stations.

Not only are these pieces of the boat—there will be eighteen double-sawn frames, nine identical pairs—time-consuming to construct; they also add considerable weight to the boat, which can affect its performance. Few lobsterboats had double-sawn frames: a yard wouldn't expend the time and money on a humble workboat and would simply bend in all the frames. But Nat wants the added strength. Sawn frames aren't necessarily stronger in and of themselves than steam-bent frames; rather, their strength is in a different form. It can be argued that steam-bent frames are superior to sawn frames because their grain is continuous, and a single length of wood with continuous grain is always stronger than a composite one consisting partly of short grain. But the fact of the matter is that bent frames can break (often, Nat explains, because a boat is planked too tightly with mahogany, which won't compress much; when the planks expand as they absorb water, they put too much stress on the bent frames). Given that, and given that there are some very hard turns in this hull, they have opted to provide it with both types of strength. Furthermore, he likes the fact that in addition to being structural pieces of the boat, the frames serve as molds for the ribbands, obviating the need to construct separate molds.

The patterns for the frames are among the most complex on the project because they must account for a changing bevel, making them in effect patterns in three dimensions rather than two. The actual frames will be constructions: two layers of 1-inch stock, locust or angelique, will be bolted together, the pattern will be traced on them, and they'll be sawn out. They'll look a little like gigantic boomerangs with varying degrees of bend. The frames for the first stations, those toward the bow, will have only a gentle sweep to them. That sweep will increase steadily as you move amidships, as in a sailboat, to form the displacement half of the hull. But continuing on, the frames will eventually bend 90 degrees to shape the flat aft section.

The sides of these frames will be beveled to accommodate the planks that will bend around them, with the deepest bevel in the forward frames, where the planks will converge hard on the stem. The bevel, furthermore, will roll along each frame, so the top of the station-two frame, for example, may be beveled only 1 degree, whereas at the waterline of that same frame the angle may be as steep as 17 degrees. Each piece will require two people to cut it out on the ship's saw—one to guide it through the blade along the traced pencil line, the other to read the degree of bevel marked along the wood and to crank the blade to the designated angle as the piece is sawn out.

The width of these frames will vary, too. They'll be about 3 inches at their heel, where they'll be bolted to the floor timbers, and will taper to about 1½ inches at the sheer, where less strength will be needed. Each pair of frames will be a kind of abstract sculpture, each completely unique.

Bruce Davies and Ted Okie, one of the few *Rebecca* crew members brought over to the shop after the shutdown, begin to lift the patterns for the frames and transfer them to plywood. Bruce has the weathered complexion and shaggy gray hair and mustache that make guessing his age with any certainty impossible. He could be a very fit sixty-year-old or a haggard forty-five—he's simply spent too many years of his life in weather, much of it tropical, and always around boats, to reveal his chronological age on sight. In fact born in Rhode Island in 1950, he joined the Coast Guard after high school in 1967 and spent three years, eight months, and fifteen days in service on *Eagle,* a 295-foot bark, a square-rigger used as a training boat. After this service he alternated between working *on* boats (doing repair and construction) and working *aboard* boats (as captain, pilot, and, for a few years on the Vineyard, able-bodied seaman, deckhand, and quartermaster for the Steamship Authority)—but always on or around them.

Bruce is a puzzle. Part of him is literary and grand, part of him is sloppy and lazy. He's reluctant to come to work, and quick to depart. He collects first editions of Edgar Rice Burroughs, Arthur Conan Doyle, and Mark Twain. He is proud to be a part of the powerboat crew and cites it as an example of the superiority of wooden boats. "With this boat," he says, "all the pieces are gathered from all over the world and put together by artists. With a 'glass boat, it comes in a big barrel from New Jersey. . . . The value of a wooden boat goes up every year like a house's. The value of a plastic boat goes down every year like a car's." As proof of this he notes that his own 40-foot schooner, *Estrela,* built in Boothbay in 1975 for $16,000, was totaled by Hurricane Hugo in St. Thomas. Bruce bought her, rebuilt her, and believes he could sell her today for as much as $100,000. Another boat similar to his, but made of fiberglass, was likewise damaged in that same storm; it sank and was simply abandoned, remaining visible for years just below the surface.

In the winter Bruce lives in a downstairs room of a house he rents with Jim Bresson in Vineyard Haven, content with his first editions, a small TV, a VCR, and his cigarettes.

✳

Nat has returned to the boatyard. It's been so long since he's been a daily part of this yard that his presence feels unusual. He begins work on the main keel timber, the biggest single piece of *Elisa Lee,* now measuring more than 30 feet long, 2 feet wide, and more than 8 inches thick. He lifts a pattern off the lofting floor onto ¼-inch plywood, many pieces of it screwed together end to end. He brings this long, floppy pattern outside to the timber resting on large wooden blocks and lays it on top to trace its outline. He then stands back to think about it.

Ross stops to regard the timber with Nat. The previous day's snow has passed, and the sun is bright, the sky a clear, deep blue.

"Might be easier to do with a chain saw," Nat says.

Ross agrees: "Chain saw works great."

"I'll need lines on both sides."

Ross nods. "It's a chore to drill." They continue to stare for a moment longer, and then Ross says, "I think I'll take my pattern and go up to Bargain Acres, see if there's anything there."

Several people are marshaled to flip the timber so Nat can work the other side. The piece, weighing more than a couple of thousand pounds, is easily flipped with levers. David Stimson, now at work on his own boat, which he's hauled out here, has stopped to help. Nat holds his ruler to various parts of the timber, shakes his head, and says to David, "You get these massive pieces of timber, you think you can make anything with 'em. And you still end up juggling quarter inches."

David says, "Oh, I see, depth."

"It'll be all right," Nat says. "We'll scarf a worm shoe onto it. It'll be better. If it gets banged up, it won't be an integral piece. It'll be easier to repair. And he'll bang it up—you know how those musicians are."

"Watch what you say," says David, a musician himself.

Nat chuckles, and David returns to his boat. Nat will take this enormous timber down to its proper dimensions with a chain saw, Skilsaw, then remove an inch off its width with an adze over the next several days.

✳

On clear afternoons the low sun angles into the shop, alighting on wood grain and sawdust. At around three-thirty the shop turns from dark brown to a bright honey-gold. The grain of the angelique Jim is carving lights up as if a switch had been hit on some inner bulb. The old tools and machinery, the chisels and planes, the work surfaces, all covered with wood shavings, golden excelsior, seem to glow. During the hour that this winter sun angles in, the shop appears to be a strange, magical room. Soon after that, dusk seeps in and lightbulbs are switched on. Heat leaves the building with the sun, and breath once again is visible in clouds of exhalation, smoky in the electric light. Work carries on till after six each night. By Friday, the end of the first week of work on the new boat, most of the pieces for her backbone have been cut and are being worked by hand into shape. Patterns for the double-sawn frames have been made, along with patterns for the floor timbers, triangular pieces to which the frames will be bolted, and the first of these have been cut out. The fundamental pieces of the boat—stem, keel, frames, transom—are coming into being.

III

Monday is clear and frigid, the cedar-sided boatshop crisp against a deep-blue morning sky. Duane, who works alone on the Bella off the outer shop, prods the fire in the stove, lighting shavings and feeding it scraps from a large bin beside the stove. Hand tools that are too cold to handle in the morning can be set here until they're warm enough not to freeze to the skin, and cans of roofing cement will be hung over the stove to soften in the frigid temperatures of February. Jim sweeps sawdust into huge piles. A new stack of wood sits outside the shop in the dirt-and-gravel lot, a delivery from Jim Aaron that came at four yesterday afternoon and was unloaded by Aaron and Ross—locust and oak from western Massachusetts. Ross arrives this morning shortly after eight. Jim, Ted, Bruce, and Duane stand around him waiting for directions. Ross tells Bruce that he pulled out potential crooks—where the wood grain curves—from the locust as they were unloading; Bruce should go through them with the sawn-frame patterns, looking for matches. There are some severe curves at *Elisa Lee*'s aft stations, and Ross is after pieces of wood whose grain has an identical curve. "We want the grain to grow right out of that frame," he explains to Bruce, who nods. "Use locust if you can, it's the closest to angelique. Don't make do. If it's not good enough, we'll get you what you need." These frames, with their grain, are some of the most important pieces of the boat.

Ross instructs Ted to move the oak that's stacked outside the shop into the woodshed—a red wooden structure directly behind the shop—sorting it as he goes by type: "Good oak—framing oak, steaming oak—and shitty oak," Ross says. To Duane, he says, "The oak marked as 'steaming' was *gray*. It had 'bending oak' *written* on it. It was dry as popcorn." Not ideal, that is, for bending along the inside of a boat hull. The clutch of boatwrights moves outside the shop as Ross says, "Let's get Nat's keel timber on sawhorses."

Now that Nat and Ted have adzed an inch off its surface, the main keel timber is close to its ultimate dimensions—7 inches wide and 22 inches tall at its largest points, and 30 feet long. The one-ton timber must now be moved from its spot outside, around the corner between the lead smelter and the steamer, and into the shop, to be placed on blocks above its pattern on the lofting floor. With nothing more than some rollers and levers, five boatbuilders complete this task in twenty minutes. There is something energizing about beginning the work-week this way—moving a massive boat timber—in part because it's a feat that really shouldn't be possible in this age of motorized assistance, and yet it is easily accomplished by flipping the timber onto rollers and levering it forward. Moreover, when the main keel timber is in place, it feels to everyone as if work on the boat has truly begun: there it is, the first piece, the boat has started. When it's fixed on the blocks a few feet above its image, Ross picks up a pattern he's been working on and says, "This is the last piece of the backbone," and then he heads into the shop to saw it out. With virtually every part of the keel made, and the first timber in place, they can begin putting the boat together.

Jim fastens his pieces together first, clamping the knee to the fore-keel and, with a drill bit nearly 1½ feet long, boring a hole through the two pieces. He tries to drill straight, but the angelique is hard, and drilling it requires a lot of pressure. The drill bit bends. When the bit at last emerges from the other side, he's slightly off center in the 5-inch-wide piece. He shakes his head. He tries again. The next hole is even farther off center, and so, concerned, he asks Ross to drill the third hole. Ross does, and *his* comes out off center.

Jim, feeling slightly vindicated, says, "Maybe eets za grain of za wood."

Ross responds, "Maybe we're just drilling them crooked."

The holes aren't dangerously off center—they'll do—but they're not perfect. Jim unclamps the pieces, smears the surfaces with tarlike roofing cement, and reclamps them.

Ross has made the first bolts, in lengths between 12 and 15 inches, out of ⅜-inch bronze rods, cutting their threads using the metal lathe and a die. He fabricates the heads for the bolts out of an old, oxidized propeller shaft that's been lying around, first lathing the corroded exterior off so that it's uniform and bright as a new penny, then drilling a hole through its center, cutting it into disks, and finally pushing a thread-cutting tap through each disk. Two dozen ⅜- and ½-inch-wide bronze keel bolts and their heads, some as long as 3 feet, will be fashioned on this lathe to hold the keel pieces together.

The heads are countersunk into the angelique; 1-inch-thick plugs will be glued in over them. Jim wraps cotton caulk around the bolt head and then pounds it through the hole, countersinking it with a spike. The cotton will help prevent corrosion by absorbing any moisture that may seep in. Ross has been careful not to cut excess thread near the head, since exposed thread increases the surface area, and the more surface area, the more potential corrosion. Corrosion won't be a problem for the first twenty or thirty years of a boat's life, Ross says, but after that it can be if it's not prevented at this point.

When Jim has the three pieces bolted together in their final form, a big stem and forekeel, with a 90-degree bend, he must finish chiseling the rabbet, the V-shaped groove that rolls across the length where hull and keel connect. When planking begins, the bottommost plank, the garboard, will be fastened into this groove. Jim continually checks his progress with a small block of wood ⅞ inch thick, the thickness of the planking.

Nat walks through the shop and stops to watch Jim work. Jim chisels carefully, slowly.

"Cut a pocket out here and here and here," Nat says, pointing to three spots on the rabbet outline, about a foot apart. "Then connect them. For some reason it's easier to read."

Jim nods. He considered this but he didn't trust himself. He soon develops a rhythm and a feel for the rabbet. Every now and then he takes a break, holding the chisel and mallet in his left hand and stepping back to regard his work. He draws his fingers through the groove in the dark-brown wood and says, "Eet's a nice curve, eh?"

He continues to work and takes it through the sweep almost to the end of the forekeel, which will be attached to the main keel timber. He'll wait till it's all together before finishing it. Ross stops beside Jim to look at his work. "Oh, nice rabbet," Ross says. He reflects for a moment. "That's a beautiful thing."

✴

When Jim has completed the rabbet, he and Nat attach the stem construction to the keel timber. The stem requires two people to carry it, but a single sliding-bar clamp can hold it up on the keel timber. Soon they have the new stem set up and fastened to the keel, and the first prominent line of the boat has been created, the long, straight keel tapering slightly into a tall, plumb stem. From the side, half of *Elisa Lee*'s profile has been determined. Stand before this and you can imagine a boat coming at you.

They proceed to prepare the pieces for fastening. Nat stands at the back of the keel timber and says, "This timber has sort of a twist in it." He then walks to the stem to check its fit. He gives the stem a couple of shots with a wooden mallet to center the piece. "We can draw our center lines," he says. "We'll cut this off and fair it here first, a little bit, with the Skilsaw." Jim works on the vertical faces that connect, pushing them together as tightly as possible and then taking a crosscut saw to the seam and sawing through the crack to make them flush, a technique known as kerfing the ends of a scarf. Nat and Jim will spend two and a half days more chiseling out the rabbet along the entire length of the keel, finishing their work with a tool called a rabbet plane.

✴

The bosses—"the bearded ones," as Ginny refers to them collectively—are continually beset by friends stopping by, sometimes to ask advice about their boats but more likely just wandering through, wanting to watch and chat. One afternoon Dan Adams wanders in from the beach. Nat is up on top of the keel, drilling ½-inch keel bolt holes through 3 feet of keel timbers.

It's the first anyone has seen of him since *Rebecca* shut down. Dan is of average height and corpulent, covering with a ball cap a bald scalp and a horseshoe of short, dark hair, his blue eyes narrowly set in a jowly, oval face. He regards the keel coming together, Nat standing atop it with a drill. "So you've become a lobsterboat builder now," Dan says.

Nat's teeth set in a half-smile, and he answers, "Hey, I'd love to get back to work on the schooner. You know when that'll happen." He hops down off the keel, and he and Dan speak quietly, the keel between them. Dan leaves shortly thereafter.

Nat says Dan is "kind of pathetic . . . I feel sorry for him." The man continues to shrug off the *Rebecca* situation.

✳

Meanwhile, Bruce, Ted, and now Bob Osleeb have been bolting together sawn frames and cutting them out on the ship's saw, having found as many crooks of locust as possible for frames with big turns in them. They've also been cutting out angelique floor timbers, triangular pieces notched to fit over the keel and beveled to match the attaching frames. When these pieces are completed and all the bolts have been cut on the lathe—two for each floor timber—then the frames can start to go on. Holes are drilled one by one through the keel pieces, smoke and damp sawdust erupting out of each as the drill works its way through the angelique; every scarf, every attaching surface, is slathered with warm roofing cement; long bronze bolts are sledgehammered through floor timbers and keel pieces; nuts are ratcheted tight on the other end. The frames are bolted to the floor timbers almost as soon as they're made.

Nat drills many of the holes, first with a short bit, then with a long barefoot auger that bends when he puts his weight on it. He must repeatedly pull the bit out, bang off the clots of wood, and then continue to drill to the other side. Bolts are then sledgehammered upward through these holes, lengths of cotton caulk having been wrapped around their heads before they're driven home. Before these U-shaped frame-and-timber structures are set up, 1-by-6-inch boards are nailed to their tops, athwartships at the sheer, to steady and support them. All work efficiently together, and the frames rise one after the other, so quickly that the boat seems to be inflating like a hot-air balloon.

Jim stops to step back and look. For the first time the shape of the boat can be visualized along the guides of the sawn frames. "That's one of the most beautiful moments of boatbuilding," he says. "After that," he adds solemnly, "it's not so beautiful." Then he laughs. People who enter the shop, and other G&B workers who pass through (such as Chris Mullen and Robert Bennett, repairing *Liberty* outside the building), halt, stare, and shake their heads, grinning. With darkness coming on fast, the next-to-last frames and timbers are bolted in. Just two weeks after the first pieces of the puzzle were cut out, *Elisa Lee* is starting to look like a boat.

✹

Ross intends to flip this skeleton upside down, fair the frames, bend in the oak frames, plank her up, caulk and fair the hull, and then flip her right side up again. He thinks it will make the work much easier, especially at the back of the hull, which is practically flat. Nat, though, feels it may be more trouble and harder on the boat to flip her twice, the second time when she weighs the majority of her anticipated 14,000 pounds. Nat says he's leaning toward building her right side up. She's just too heavy a boat, and she takes up so much space as it is that there's scarcely enough room to move around her now—if they flip her upside down, the widest part of the boat will be right along the floor.

But Ross says, "I just can't imagine planking, fairing, and caulking her right side up. Half those planks are on the bottom"—he makes a curving right angle with his hand—"and my knees and shoulders just don't work that way anymore."

Ultimately, Nat stops playing devil's advocate, and he and Ross agree that it is best to build the boat upside down.

IV

Monday-morning gusts blow snow swirling into the shop when the door slides open and Jim and Ted and then Bruce enter, each leaning with a shoulder to close the door behind him. The plastic stapled to the frames extending off the outer shop, a temporary shelter in which Duane is completing the Bella, has blown loose and is popping loudly in the wind. The temperature hits 12 degrees by 8:00 A.M., but the thermometer outside Ginny's window won't rise above 18 degrees all day. Ted, in a heavy green jacket and a Depression-era wool cap, lights a fire and soon has it vigorously stoked and crackling. Jim, wearing his perpetual jeans, bed head, and sweat-shirt beneath a filthy beige down vest, stands near the heat, as does Bruce. Holes in the knees of Bruce's jeans reveal long underwear, critical on a day like today. They all stamp their feet and hug themselves in the cold.

By ten after eight, neither Ross nor Nat has arrived, and Jim nods seriously and says, "Za boss has been making party yesterday." Then he chuckles, but his amusement fades beneath the weight of the cold. "This is heavy starting Monday," he says. "The weather doesn't make you feel like you want to work. Maybe we should just go home and drink some grogs." He grins, nods quickly, and begins to laugh. He has worked all weekend on his boat, he explains, putting Dynel, a durable cloth-and-epoxy coating, over his plywood deck.

With neither boss in evidence and no work to lessen the effect of the cold, Bruce is gone—it's too cold to work today. Bob has headed "to America," he said—to Boston—to obtain a Massachusetts driver's license, though neither he nor Marilyn owns a car (they bike to work together all winter).

Nat arrives before Ross, at around eight-thirty, wearing his beret and green hooded sweatshirt and carrying his canvas bag, tubes of rolled drawings in it, for a schooner he's creating for his friend Ormonde. Nat clears his throat, says good morning to all, and heads to the office. He and Pam are leaving soon for a short vacation on Cumberland Island, Georgia, where they'll stay in a house loaned to them by friends. (Nat and Ross each used to spend as much as two months away from the Vineyard every winter, usually sailing in the Caribbean, but with increased work, such breaks have become less and less frequent.)

The "Pepe's" truck rattles into the parking lot shortly behind Nat, and once Ross comes in, work begins quickly. The final frames are bolted in and securely strapped with ribbands. And then, at last, the boat is rolled.

The idea behind flipping a long, heavy object is simple: lash her backbone with heavy rope and hook that rope with shackles to a block-and-tackle, a two-to-one purchase in this case, bound to the collarties above, and heave-ho—she flips of her own weight. Ross takes one line and Ted the other, and each pulls until the keel shudders on the sawhorses and the floor of the shop rumbles, and it's clear then that this is a very heavy object indeed. The main keel timber weighed nearly a ton when it was rolled into the shop and lifted onto sawhorses; now, with all the additional keel pieces, floor timbers, and frames attached to it by bronze bolts, it weighs almost double that. Ted and Jim begin to laugh out loud, it feels so exciting and potentially dangerous. As Ted and Ross hoist the keel, the incipient hull wants to roll, and Jim drops to the floor to try to guide the frames and keep as much weight as he can off the ends as she rolls.

This work is slow, Ross being careful with all his instructions and knots to avoid any injury to crew or boat. Bolting in the final timbers, preparing her for rolling, and then doing the actual rolling—all of it together takes a few hours.

Once she's around—keel in the air above frames, boat now upside down and dangling—she must be supported, but Ross lets her list. He stands there squeezing his chin, fingers disappearing in whiskers, and staring at what looks like a large animal. He stares at the floor and then back at the boat. He's trying to decide how best, he says, "to position the three-dimensional object over the two-dimensional drawing."

Indeed, this big boat is suspended over an exact scale drawing of herself. Where Ross positions her now, she will stay until she is planked, caulked, faired, and rolled back over. He wants her fixed mirror-perfect over the drawing, as precisely as if an underground light were being projected through the lofting-floor drawing to form a hologram of the lobsterboat hovering in the air. And so Ross positions the stem over the drawing of the stem, and aligns the frames of station one exactly along the straight pencil line that runs across the floor representing station one.

They build a structure beneath the boat for her to rest on, first sliding below her huge blocks of wood gathered from out in the sand and around the yard, then ultimately hammering together two layers of 2-by-10's into two 30-foot supports and resting them on edge beneath the port and starboard sections of the boat. Jim, Ted, and Ross toe-nail all the pieces in as they go, securing them at their base with staging nails, nails that have two heads so they can't be driven all the way in and may be removed quickly and easily. The men not only secure the boat to the 30-foot planks on blocks, but also clamp two more long planks to the hull and nail them to the boatshop wall. Ross pushes the boat fore and aft, leaning into her slow and heavy like an elephant to see how secure she is, then repeats the test from side to side. He adds more supports at either end. He marks the center line with a long string fastened to the stem and transom at the waterline. When the backbone-and-frame construction is perfectly centered and perfectly level—as if it were floating at rest in an upside-down pond—Ross makes one last check to be sure that the stations are on their lines, then finishes nailing every piece of wood to its supporting piece. Blocks of wood are nailed into the floor to secure the stem, just inches above the floor. *Elisa Lee* is fixed.

※

As the last nails are driven in, Kirsten Scott arrives at the shop with her sister Kym and Kym's young son, Farley. Kirsten has been visiting

Ross frequently, and when she's in town she often brings him lunch. They sit by the fire or, if the sun is out, on the dock, and they eat and talk.

Since that day the previous fall when she came looking for him on *When and If*, Ross has been openly and delightedly smote by this thirty-five-year-old fisherman, tall and lean with blue eyes and long brown hair. If Kirsten appears suddenly while Ross is fixing an electrical cord at a bench, he will look up with a jolt, his eyes will seem literally to brighten, and he will say, "Where did *you* come from?" and grin so hard at Kirsten that there isn't anything for her to do but laugh. Kirsten for nearly a decade fished the waters of southeast Alaska and Bristol Bay, and while there is a softness to her appearance, her handshake is designed to tell a captain that she's every bit as qualified as the next man. She fished for less than half of each year, work that paid enough to let her travel and play the rest of the year, a schedule shared by her sister. Kym, the darker one, flies a Cessna 180 on floats in Alaska and a 1946 Cessna 140 on the East Coast, spotting for the fishing boats. Kirsten and Kym are two of seven children raised in Lincolnville, Maine, by a divorced mother who supported the family by doing boat work and other odd jobs. The Scott children were known throughout town for being, quite plainly, wild, and now all but the two youngest girls worked the sea hunting wild creatures.

Kirsten is game for any adventure, has a ready smile, and loves to cook. She grows animated when she talks about fishing for king salmon, and will explain in great detail how, when you bring such a fish to the boat, you have to crack a gaff hook down hard between its eyes and in the same motion swing the hook end around into its gills to get it into the boat. If you miss with the first blow, the salmon will get angry instead of stunned, and when a salmon is mad, it can get away. "King salmon," she says, "are often the most gentle; they swim *so* sweetly up to the boat—*one good whack and he's yours.*" King salmon sell for between $2.00 and $2.50 a pound, so a good-sized fish can be worth a hundred bucks, and you don't want to lose a hundred bucks because you miss with your gaff hook. Once the fish was in the boat, Kirsten would first gill-bleed it, then gut it, always feeling excited to scoop through salmon entrails. Sometimes, she explains, she would find squid so fresh in a salmon gut that she'd clean the squid right there, too, take it below, and sauté it in a little olive oil and garlic. Her eyes narrow at the thought, and she says, "They're so *fresh*—they're *delicious*."

On this frigid, blowing day, she and Kym have brought a pot of borscht and a couple big loaves of crusty bread, along with bowls and silverware—more than enough for everyone. She has set the soup atop the wood-burning stove, and it's piping hot. She cuts big chunks of bread with a knife, using the ship's saw as her board. The wind is blowing out of the north, hard, and the schooners in the harbor pitch in the waves; walk out on the beach, and icy sand whips your face. In the cold, dark enclosure of the shop, the boatwrights sit on stepladders and crates and machinery to eat their hot soup and fresh, dense bread. They are mainly quiet, letting the soup warm them. Anyone who enters the shop, whether friend or stranger—and there is rarely a day when several such people don't pass through—is offered a bowl of borscht. After the good morning of work, the meal makes Jim feel so happy that he walks across the street for a coffee and a packet of cookies. When he returns, all remain seated, sated and quiet. Breaking the silence, Ross says softly, "Well, I'm gonna get back to work."

Nobody feels like working. Jim simply says *"Borscht,"* like a drunken man. All nod in agreement, having eaten too much. Jim says, "We are sitting below our waterline." He is serious, but then he throws his head back and laughs.

✷

The outer shop has been transformed: a great beast has been driven in here, one with a long spine the color of burnt umber that rolls forward into a round skull, its rib cage reaching out from its backbone to the ground. Looking at it produces the same effect as walking into the dinosaur room at a natural-history museum. The creature is nearly twice as big as a stegosaurus. Just as a reconstruction of fossilized bones invites you to visualize what a dinosaur must have looked like when it walked the earth 150 million years ago, a boat skeleton likewise causes you to imagine what *this* life-form will be like when it comes alive to ride the ocean. The overturned boat creates a new dynamic in the shop. There's magic in a skeleton because it is half realized and half imagined. The ribs are real and determine the shape; you can bend your vision around them like planks.

These frames are significant, but an understanding of them is a matter of perspective. Viewed in the context of the past several decades, their very existence here is uncommon, even remarkable. As fiberglass moved in to dominate boat construction—with monocoque hulls popped out of molds—frames began to disappear. Today, new construction using traditional wooden boat building techniques—planks wrapped around frames—is a rarity.

Indeed, frame-first construction is practically a fad if you consider that the vast majority of mankind's history of creating watercraft was spent building shell hulls—that is, shells were somehow fashioned, and if necessary, frames were inserted for extra support.

Between thirty and forty thousand years ago, humans began to lash logs together, creating platforms to carry themselves across water. Ten thousand years ago, man had the tools and the know-how to carve out big logs and make them float upright.

The next major benchmark in wooden boats came during the Bronze Age, circa 3000 B.C., with the introduction of planks to create sophisticated hulls not unlike the ones we know today. The oldest ship for which physical evidence survives was built about 2650 B.C. in Egypt—a riverboat more than 140 feet long, propelled by paddles. From this time on through four thousand more years, planks were fastened edge-to-edge by various means. The Egyptian riverboat hull was joined by mortise and tenon, while other hulls were sewn together with willow fibers and caulked with moss, or fastened with trunnels. These vessels cruised the Mediterranean, trading and fighting, fighting and trading. In all cases, frames were inserted after the hull took shape, to add strength.

Eventually a few civilizations figured out that if you overlapped the planks, you could achieve a shapely, strong, efficient hull. In one case—that of the Vikings—such construction advanced to the point where it allowed the civilization to expand from Scandinavia to Russia and North America and even as far south as the Mediterranean, during the eighth, ninth, and tenth centuries. In effect, it made them famous.

The Vikings were really the first people truly to be defined by their wooden boats—those shallow, fat vessels with the sleek profiles and scimitar-like bows. As one maritime archaeologist has noted, what the temple was to the ancient Greeks, the ship was to the Vikings. Another historian has written that the Viking ship was "one of the greatest technical achievements of northern European society before the building of the early cathedrals." The Viking ship represented the pinnacle of its civilization's technology and material culture. It was the foundation of its power, the means of its expansion in the Northern Hemisphere. These boats also remained the Vikings' greatest *artistic* achievement—vessels so beautiful and distinctive that their lines would still be popular design elements in North American societies a thousand years later (note, for example, the *WoodenBoat* logo, showing the front view of a Viking ship). The Vikings are legendary for their bellicose behavior, for the rape and pillage of every shore they landed on. They neither invented such behavior nor excelled at it beyond all others (the blood shed by Vikings over several centuries is a drop in the bucket relative to recent decades)—but they became *famous* for it because their boats were so good. A warlike nature is evidently innate in humankind; superlative vessels allowed the Vikings to better express their inner nature.

In these ships, too, planks were fastened to one another to determine the shape of the boat and form its hull, with frames being fit in later. Not until the thirteenth or fourteenth century did framing a boat *first* begin to make sense to boatbuilders. Now, for the first time, a boat had to be completely designed ahead of its building, rather than each successive plank's being allowed to determine the next. Planks were no longer fastened to one another, as they had been for four millennia, but were instead attached to the frame. This development, like most technological advances, seems to have been motivated by a practical need—in this case, to travel farther—but the exact reasons for these new ideas, and how ship builders arrived at them, are not known. Bigger boats were needed to carry more people and more cargo for longer periods; shell-built boats could be only so heavy, and therefore only so strong. By the late 1400s, the big sailing ship with a three-masted squaresail rig was established, a technical achievement that would change the world.

Planks attached to frames allowed for heavier, stronger vessels that could cross oceans, last months at sea, and carry enough guns, food, and people to dominate new lands, not unlike the Viking ships but on a vaster scale. These sailing ships were easier to maintain in that they didn't require frequent repair. They were the vehicle for all-but-global European predominance. They changed mankind's view of the world (notably when Captain James Cook charted much of the Pacific in the mid-1700s) and altered its actual complexion (with the creation of the United States of America, for example). The next major innovation came in the form of rigging, as the fore-and-aft rig replaced square sails for easier, faster, more maneuverable sailing. As to hulls, the plank-on-frame boat would remain dominant until cheap iron plate for iron hulls became available in the mid-nineteenth century.

The introduction of frames into hulls, with all that resulted, is considered by many equal to the harnessing of steam power or electricity, producing revolutionary changes in the history of human civilization. It would seem difficult to overestimate the importance of the plank-on-frame wooden boat. In many ways—economic, political, social—it shaped the foundation of Western society.

✳

The construction of the schooner *Rebecca* and, except for the obvious differences in design to account for her engine, of *Elisa Lee* scarcely differs from boat construction of hundreds of years ago. Wood still behaves the way it did in the Bronze Age. The main difference today is in the tools—power tools make an extraordinary difference, change the work, and therefore change the people who do the work.

The British nautical historian Basil Greenhill describes the life of the boatwright in the late nineteenth century, considered the golden age of plank-on-frame sailing vessels, in his book *The Evolution of the Wooden Ship*. Life was harder then; any modern-day American would find it ridiculously painful in comparison with the ease and comfort now enjoyed.

Workers often lived many miles from the shipyard and started off their day with a five-mile walk to work, which began at 6:00 A.M. and ended at 6:00 P.M., followed by a five-mile walk home. They were given a break on Saturdays, when the day ended at 4:00 P.M.—a ten-hour stretch rather than a twelve-hour one. In wet and cold weather, their flannel shirts and wool trousers did not dry out overnight—and waterproof clothing hadn't yet been invented—so they spent most of the winter in wet clothing. Sunday was the only day of rest, the one day a shipwright had free to labor on his own boat. In slim off-hours, at the public houses, boatwrights talked about their work. These men devoted most of their waking hours to boatbuilding from their first apprentice days—part-time from age seven and full-time from age twelve or thirteen—and most of their off-hours talking about it, so they became very, very good at the work; the boatwrights of the time were men of extraordinary technical skill. They spent much of their day just sharpening tools, which were, needless to say, kept very sharp.

Boats were first designed in half models, layers of pine pegged together and carved to the desired shape. Such a boat hull could be judged not only by eye but also by touch, the fingertips picking up small imperfections that, on a finished hull, would no longer be so small. After this scrutiny, the wooden pins of the half model were removed, and its layers were taken apart and either traced to create a lines drawing or simply measured with a 2-foot rule and transferred directly to the lofting floor. Nat's first designs were half models such as these, but now he starts out straightaway with pencil and paper; Myles Thurlow, who apprentices at G&B as part of his Charter School curriculum, in his own right a fine young sailor of fifteen, is crafting the half model for the boat he intends to build. It remains an excellent way to design a hull.

Once the boat had been lofted, the timbers were sawn—big keel timbers, frames, hull planks, and thin planks that would be used to make patterns. The typical method for sawing a large tree into planks called for the log to be rolled over a pit, with one sawyer standing on the log, guiding the saw, and a pit boy down in a hole of the same dimensions as a grave, heaving down on the pit saw. The saw blades were eight feet long, with deep gullets between their teeth to carry the sawdust out. A good hard pull on one of these pit saws could move through three inches of log. Sawdust rained endlessly onto the pit boy, making for miserable work for which the boy earned as much as three and a half shillings a week—in six weeks he earned just over a pound.

Keel timbers and frames were sawn the same way. The interior of a sawn frame was too curved to saw and so was cut with ax and adze.

Greenhill notes that because it was so difficult to find big pieces of curved wood, the ship builder was forced to construct them out of several pieces "joined together in such a manner as to allow the run of their grain to follow the curvature of the frame as closely as possible." The keel pieces, stem post, and sternpost were bolted with iron more than 1 inch in diameter, pounded through holes that had been bored with cross-handle borers and crank augers. The rabbet was roughed out on separate keel pieces by adze and chisel, with the master shipwright completing "this complicated groove mostly by 'eye' from long experience," according to Greenhill.

While power tools, weather-resistant clothing, and the vacationlike hours of 8:00 A.M. to 5:00 or 6:00 P.M. made the work at G&B a spa compared with that endured at yards of the previous century, the actual construction of the skeleton of a boat remained a virtually identical process.

V

On Saturday morning Jim arrives at eight, and by ten he has made good progress on a cabin corner for his boat, all corners being critical spots requiring special attention. Jim has made the transformation from uncertain apprentice to boatwright by the slowest but surest method: he has focused relentlessly for six of the past seven years on boatbuilding. In his late twenties, unmarried and with few responsibilities, he has plenty of time to spend on boatbuilding, not to mention on reading about sailing and carousing with Ted in the evenings. Even with all this time, though, he rises at five-thirty throughout the winter and arrives at the shop by six, when the harbor is quiet and the shop is deserted—no tourists, no prospective buyers or dreamers wandering through, no Bruce and Bob and Ted to take his attention away from his most important work—intent on restoring his boat *Tomahawk* to sailing readiness by race month, September, half a year away.

Jim must reframe, replace the garboards, refasten the planking, recaulk it, put on a new deck and a new cabin, and build a new interior. He has, in effect, bought a lead ballast and an oak keel, bent frames, and planking stock, partially assembled, for $1,000. He spends five days a week working for G&B, technically as an independent contractor, from 8:00 A.M. to 6:00 P.M.—taking the hour between 1:00 and 2:00 for lunch—and then for two hours a day during the week and as many as he has energy for on the weekend, he works for himself on *Tomahawk.*

Bob comes in at ten and hitches one strap of his Carhartt overalls over his shoulder. He's missed a few days of work recently and needs to put in a Saturday to saw and bevel floor timbers, the trapezoidal pieces of angelique that are notched at their short edge and fitted into the keel at each set of frames. The floor, or sole, will be built on top of these after the boat is righted.

Bob transfers his floor timber pattern to a heavy slab of angelique and saws the pattern out using a Skilsaw. He then tightens the piece into a vise and begins to plane the bevel.

Jim is still working in the shop, now at the table saw. He has removed the metal guide that runs parallel to the blade and in its stead has clamped a piece of wood diagonally to the blade. This will serve as a jig, or guide, for the cabin corners; he wants the inside of each corner to be curved, not square, and so will cut a semicircular hollow about 1½ inches wide through a long piece of angelique. Typically you push a piece of wood straight through the whirring circular blade of a table saw, simply cutting it into two pieces. But if you lower the blade so that it's an inch high, and you push a 2-inch-thick piece of wood over it at, say, a 5-degree angle, the blade will cut a tunnel through it.

"Ross just showed me this last week," Jim says to Bob.

"You didn't know that? I'm surprised. Because you know so much. For a young guy." Bob chuckles. He loves the table saw and all that you can do with it, all the attachments you can hook up. "They don't use table saws nearly as much as they could. There's lots you can do with a table saw." Then again, he notes, G&B's notion of tools generally is unlike that of anywhere else he's worked. "They use band saws here," he says. "Look around, we've got four band saws."

Jim nods and says he likes G&B's uncommon relationship with tools. "Eet learns you how to work with your eyes," he says. He looks around the shop at the ancient cast-iron machinery that never works exactly right but works right enough for what these boatbuilders do. "It could be all swept up, with new tools," he says. "It wouldn't be the same."

Bob says, "If you can work here, you can work anywhere."

Bob is a reader, tending toward books about sailing and boatbuilding. As he takes fractions of inches off a dark-brown floor timber with a plane, he talks of his current book, a biography of Joshua Slocum, who became the first man to complete a solo circumnavigation when he cruised into Newport, Rhode Island, in June 1898 on his 37-foot yawl, *Spray*.

Apparently, the practice of giving away derelict wooden boats to delusional sailors goes back at least to 1892, when an acquaintance of Slocum's, a whaling captain, offered him a ship if he'd travel to Fairhaven, just across Vineyard Sound, to fix her up. "Next in attractiveness, after sea-faring," Slocum felt, "came ship-building." And since he could not find gainful employment as a captain or a shipwright in Boston, he decided to get to work for himself in the latter capacity in order to pursue the former. What he found, of course, was that no one in his right mind gave a decent wooden boat away (Nat Benjamin could have told you that as he refastened yellow pine planking to his 37-foot sloop, *Venture*). Slocum discovered that *Spray* had been rotting away in a field beneath a canvas tarp for seven summers. The neighbors said she'd been built in the Year One; when they met Slocum, they assumed he planned to tear her apart. When he informed them otherwise, Slocum noted, "great was their amazement."

It would take some effort, of course, to daunt a man who would be the first to sail around the world alone. Slocum simply got to work—picked up his ax, hunted a nearby oak, and cut it down. He paid one Farmer Howard to haul it to the field where his boat was set up. The tree would make a fine keel. He took down more timbers, straight saplings of oak, rigged a steam box over a pot and got the timbers cooking, then bent them over a log to dry till they set, working at the same time on cutting out keel pieces.

"Something tangible appeared every day to show for my labor, and the neighbors made the work sociable," he said. "It was a great day in the *Spray* shipyard when her new stem was set up and fastened to the new keel." He was well pleased with the sweep he'd found for his stem: "The much-esteemed stem-piece was from the butt of the smartest kind of a pasture oak," he reported. "It afterward split a coral patch in two at the Keeling Islands, and did not receive a blemish. Better timber for a ship than pasture white oak never grew."

Slocum, as the people of Fairhaven began to realize, was transforming a rotting hulk into something remarkable, and whaling captains from neighboring New Bedford, whaling central, eventually wandered over to have a look. Slocum enjoyed the attention. "When a whaling captain hove in sight," he said, "I just rested on my adz awhile and 'gammed' with him."

Slocum replanked the vessel with 1½-inch-thick yellow pine boards and fastened them with a thousand bolts, nuts tightening them to the timbers. He drove a layer of oakum followed by a layer of cotton caulk into her beveled seams. A deck of white pine was spiked to yellow pine deck beams 6 inches square, running athwartships as wide as 14 feet at the boat's widest station. Two coats of copper paint were slapped on when the caulking was finished, and two coats of white lead on the topsides, and then Slocum installed a rudder and launched her the next day, after thirteen months of labor and an expense of $553.62.

"As she rode at her ancient, rust-eaten anchor," Slocum wrote, "she sat on the water like a swan."

Five years later Slocum would complete one of the greatest sea journeys ever recorded, having been chased by Moorish pirates off Gibraltar, submerged by a wave off Patagonia, and attacked by South American "savages" while at anchor in the Strait of Magellan (he scattered tacks across his decks nightly and in the morning swept them up), having weathered countless storms, seen ghosts at his helm in times of delirium and sickness, and eaten the flying fish that *Spray*'s deck caught during a night's sailing. Slocum was a brilliant entertainer, as his classic sea story, *Sailing Alone around the World*, proves. Even in port he could drum up excitement (and money for Customs and provisions) by, say, catching a 12½-foot shark (as he did while anchored in Melbourne) and charging the public an admission of sixpence to view the creature.

Bob and Jim love these stories. "That's what keeps us riveted to the work," Bob says. He explains that the book he's reading was written by Slocum's son, and that he's currently watching Slocum saw tropical wood in Brazil. Slocum was a professional captain his whole life and once, shipwrecked in Brazil, built a new boat from scratch to sail home. "Some of those tropical hardwoods," Bob says, "it was a chore just to square them off. One guy on top, one guy on the bottom. How can we complain about anything?" Bob takes a couple more sweeps with his plane. "He spent time on Martha's Vineyard. He wasn't from here, but maybe he retired here."

"Where did he die?" Jim asks.

"Maybe here."

"We should find his tombstone. That would be neat. We could make some money." Jim suggests that they build little replicas of *Spray* and set up a stand right there on the old sailor's grave. Jim guffaws at the thought. But if the Black Dog can sell T-shirts and coffee mugs to tourists, why not?

The two boatwrights pursue their work quietly after that. Bob will go on to read in his book that Slocum died, appropriately, at sea, lost, and that he in fact departed on that fateful journey from these very shores.

✳

After another hour or so of work, Jim leaves to buy some lunch down the street at the A&P, returning with turkey, tortillas, cheese, and hummus. Bob runs across the street for boiling water to pour into his instant soup, which he'll eat with the crackers and cookies he sets behind him on the ship's saw.

I came to the shop this day because I knew they'd be working. During the week, questions and conversation are difficult; today I'd be able to engage them in more leisurely talk about boats.

Jim throws more scraps of wood into the stove, leaves its door open to enjoy the heat and bright glow, and pulls a crate near.

Jim likes building wooden boats, he says, "because they're the best. They're better in rough weather." But he adds that there are "less and less people who know the sea. And most people grow up with plastic boats.

"When you build a boat," Jim says, "that is who you are."

Bob stirs his soup but leaves the spoon in the cup, does not take a bite. "The fact that you can build a whole boat out of wood," he says, "something that was living, with a minimum amount of processing, I think that's the key. In order to build a fiberglass boat, you need fiberglass, you need fiberglass resin, you need a lot of chemical processing. To build a wooden boat, you need a tree and maybe a chisel. They built them without screws, they used trunnels, wooden pegs, they used to lash 'em together and stuff. You don't need a whole lot to build a wooden boat."

Bob takes a couple spoonfuls of soup, steaming in the cold air, and reaches for some crackers behind him on the ship's saw. "Sailing on a wooden boat is a total experience—owning and sailing on a wooden boat," he says. "You notice that people are always coming up to you. People like wooden boats. And therefore they like you. With my boat, it's very easy for me to meet people, people come up to me all the time.

"Are they better than fiberglass boats?" he goes on. "Depends on the design. Fiberglass boats sail all over the place. In fiberglass you can do more things as far as radical design, whereas with wood you have certain parameters that you have to deal with. Now, cold-molded boats, if you want to call them wooden boats, they can do things like fiberglass boats can. They're kind of like half-breeds, kind of in between.

"But one reason I chose to build cold-molded boats is because I didn't have a Brad. I didn't know where I could get good lumber from, so it's a good way of dealing with that.

"Wood as an engineering material is excellent—this is better." He nods in the direction of the outer shop. "Because you can recycle them, it doesn't really damage the environment, and that's a big reason why I like them. We're going to have to revert back to these kinds of things and live these kinds of ways and take care of our forests. Boatbuilders will take care of their forests because they know they need wood to build a boat."

Bob came to wooden boats in 1976, when he was twenty-three. He'd drifted down to Key West and bought a 30-foot sloop, *Givin,* and there he began building and repairing wooden boats. And it was with this boat that he began to learn something about the nature of wooden boats. It happened in a storm, and it's left its mark on Bob, as is clear from his hushed tone as he tells the story. But he won't make any leap out loud—*Make up your own mind,* he seems to say. He keeps his lips clamped shut and shakes his head. I have to prod him a little.

It's often noted among wooden boat people that their boats have soul. This utterance is so common that by now it's almost a cliché. Often people claim, freely and openly, that wooden boats have *a* soul—that is, their own individual souls. But when Bob finishes his story, his whole clenched body says, *You're not going to get* me *to say that*

wooden boats are literally *alive.* But his very resistance implies otherwise:
How else could you explain it? Chance?

"I don't want to get into it now," he says, staring into the fire. But
he does, of course. Because wooden boats always come with stories.

✳

As dramatic weather often does, it came upon them quickly, he
says. It got bad fast. Bob, his first wife, Cathy, and their dog Coco had
been preparing to sail for the Bahamas that day, in April 1979. *Givin*
was anchored outside the harbor near Christmas Tree Island with some
other boats, a quarter mile off Key West. It was not a good anchorage
because there was no protection from the wind. In Key West there's a
small harbor, with a narrow opening—Bob remembers it as being
about fifty feet wide—that's big enough for only about ten boats.
Today the harbor is overlooked by Mallery Square and a nightly carni-
val at sunset, and it's filled with charter boats, but back then it sheltered
working fishing boats. Bob had taken *Givin* off her mooring in preparation for their departure later in the morning. He and Cathy and
Coco had gone to the store, and the two humans hugged big bags of
groceries to the dinghy; they hopped in and pushed away from Sim-
ington Beach, and Bob began to row out to *Givin.* He felt the rowing
grow a little harder—strange, since it was a fair day. Then he looked
north, and his stomach dropped. A storm was right there, had come
out of nowhere, with no warning whatever, straight at them, and fast.
That wasn't rain in the darkness, Bob saw; that was the ocean being
lifted straight up into the air. They'd rowed about fifty feet. Bob heaved
on the oars hard a few times toward *Givin*—she had only one anchor
out. She was in danger and would endanger other boats. But he
quickly realized he had no chance of reaching her in time. In fact, he
worried about even making it back to shore, since winds were suddenly on them and they were being blown south. He rowed hard, and
they landed at the small harbor near Mallery Square, about half a mile
down from where they'd pushed off. He got Cathy and Coco ashore
and they found refuge in a shrimp boat. Sixty-knot winds and driving
rain were upon them, everyone scrambling to secure lines.

Bob ran to the end of the wharf, terrified for *Givin.* He shielded his eyes against the wind and rain. Between gusts, there was enough visibility to see the worst. Gone? He strained for vision—and accepted what had happened. *Givin* was nowhere to be seen.

No, wrong: there she was, heading fast downwind. A boat with an anchor out, even one not holding, will point into the wind. Someone had cut her loose. *Givin* was his house and his home, where he and Cathy lived; she contained everything that mattered to them. And there she was, being driven away from him. The freak storm had taken everyone by surprise—hell had broken loose, Bob said. He ran for the Coast Guard station but made slow progress because the streets were flooded and there was water up to his knees. Bob begged the Coast Guard for help in getting to his boat, but they said they couldn't do anything until the storm passed. Bob returned to the shrimp boat to be with Cathy and Coco and the shrimper. From the cabin he saw a boat, a 50-footer, struggling to reach the narrow opening of the harbor; it was as if she were maneuvering under a bridge. With *Givin* gone, there was nothing for Bob to do but offer help to whoever needed it, and he ran out into the storm. If the man in the 50-footer could get a line to him, he could help maneuver the boat through the narrow passage and keep her from smashing into the wharf. The man threw him a line. Bob wiped the rain out of his face and caught it, held it fast. Then he turned. He heard something in the water behind him, he turned at something knocking against the wall. Immediately behind him inside this tiny harbor was *Givin*. She was sitting right there in the water like a wild horse returned home, bumping against the wall as if to get Bob's attention.

He let go of the other man's lines and grabbed hold of *Givin*.

"It was such a little opening," Bob says, the memory clear and powerful for him even two decades later. "And such a little harbor, and she was right there. *The boat actually came to where I was on land.*"

Jim doesn't say anything, just nods and smiles. Bob didn't want to go into it in the first place; he doesn't even know why he brought the story up. You're not going to get him to say it. Humans have their destiny, and Nat says boats have their own lives, their own destiny. But don't expect Bob to make any crazy claims, because he won't. *(A good wooden boat will find her owner in a storm. She will travel against waves, across gale winds under bare poles, will slip through a narrow opening into the harbor where you're assisting another sailor—and wait by your side.)* No sir, not him. He's told Jim and me, he'll tell anyone, what happened, and we can make of it what we want. *(Wooden boats are literally alive.)* Nothing but good karma is what it was, he says—a shift of the wind, good luck.

Time to get back to work.

VI

Bruce Davies loiters in the outer shop before moving in to his work on a cold, still weekday morning. The door opening out onto the dock and the harbor is propped open. Bruce stares out and says, "A little fog this morning."

Duane Case looks up from his work on the Bella and says, "A *little* fog! It's *beautiful* fog. Haven't seen so much fog in months."

The entire harbor is enveloped in a great shroud of it. The three big schooners, *Shenandoah, Alabama,* and *When and If,* sit like ghosts in clouds on the flat-calm water.

The loft floor upon which *Elisa's* backbone is being set up drops off to sand, but this winter a small shelter of 2-by-4's and translucent plastic has been extended off the shop's frame (it's a constant irritation to Ginny, since her view of the great boats is cut off). The fourth Bella, to be named *Isabella,* is under Duane's hand. Planks are positioned around her a couple of feet in the air so he can climb in and out easily, and they buckle steeply when he does.

The Bella is just over 21 feet, a comely little daysailer but also big enough for a summer weekend trip to Naushon or one of the other Elizabeth Islands off the Vineyard. It's arguably Nat's most successful design, at least if judged on the quantity of orders. Everyone seems simply to like the look and feel of these sweet little gaff sloops, and Duane is content in his solitary work on this one.

Earlier in the week, Duane made the cabin trunk for the Bella out of a single piece of white oak, approximately 14 inches at its widest point, ¾ inch thick, and more than 12 feet long. The design of the boat requires this piece of wood to bend 180 degrees, in a perfect U, to form the raised sides of the cabin and contain two oval port lights.

Not only must this piece of wood bend considerably, but it must be spiled, because there are all kinds of curves going on here—"Nothing's square or straight on a boat," Duane says. The cabin top that will fit over it will have a convex curve, called camber, and the deck to which it will be bolted is also a convex curve; this means that the very center of the U will be the narrowest point, gradually widening out to the ends. To make this pattern, Duane held a piece of thin, flexible plywood along the curve of the deck and dragged his pencil along the bottom edge of the plywood where the finished piece will fasten to the deck. For the top edge of the trunk, he made measurements of its intended height at various points, taken from Nat's drawing, and connected the dots.

Duane sawed the piece out—slightly S-shaped, planed to a thickness of ¾ inch—and cooked it for an hour in the steam box. Meanwhile, on the spot where Ross lofted the little rowboat that Robert Bennett, who does much of the yard's brightwork, now has on sawhorses on the dock to paint, Duane nailed a series of plywood triangles, or gussets, into the floor in the arc he required. When the oak was hot and saturated, he carried it into the shop to the gusset frame he'd built and, beginning at the middle and working quickly outward toward each end, clamped the oak to the gussets, pulling that board as hard as he dared. In twenty minutes the entire piece was clamped in place, bent slightly beyond 180 degrees, both ends pointing at the table saw. This work was more fun to have finished than to actually do—he'd had a hard time finding a decent piece of oak for this important and visible piece of the boat, plus it had that terrific bend in it. But it seemed to take hold nicely.

When Duane unscrewed the clamps several days later, the oak straightened somewhat but retained most of its bend. He fastened the piece to the deck, driving bronze screws from underneath the deck up into the cabin trunk.

On the day of the beautiful fog, Ross arrives shortly after eight, and Duane calls to him as soon as he enters the outer shop. Duane has begun to mark the position of the port and starboard port lights on either side of the cabin trunk, and the end of the oval that will be cut for the starboard one runs directly through a knot. He saw the knot earlier but hoped it would fall right in the center of where the port light would be cut out. No such luck.

Ross stares for a moment. Knots are no good. You never know how they'll behave under stress. Ross says, "Let's move it here, back to a flatter side of the cabin trunk." He reaches for the pencil at his ear, but he has just arrived, and there is no pencil there yet. Duane hands him his, and Ross sketches in an oval where he thinks the port light should go. It will be mirrored on the port side. It is a situation in which a particular piece of wood has made its own design change. Duane agrees, and he and Ross head into the shop to look for the right oval pattern for the port lights.

Duane is thirty-nine years old and lives on a 32-foot double-ended ketch, tied up at the dock behind Maciel Marine, with his wife, Myrtle, a Scot twenty-three years his senior. Duane is of average height and build, skinny, as are most here due to a spare diet (Duane and Myrtle are fond of beans and cornbread). His straight auburn hair is thick as wool and seems less brushed than pushed or maybe even shoved into place. He wears baggy jeans, sometimes denim overalls, flannel shirts, and old, brown, hard-soled lace-up shoes. In very cold weather he'll put something on his head, though the hair goes a long way toward providing insulation, as does his shaggy beard. Duane is unique in the boatyard, in this work that requires so much visual judgment, in that he was born blind in his right eye, and now his one useful eye is vastly enlarged behind the thick, often foggy, lens of his glasses. His voice, a deep, somewhat nasal tenor, is resonant when he speaks. Because he works alone, though, this voice is rarely heard in the shop—unlike those of Bruce and Bob, who bicker all day long, Bruce continually baiting the earnest Bob. When asked a question, though, Duane will answer with articulate thoroughness. The Bella is now planked, caulked, and faired, her deck and rails fastened, and once the cabin trunk is complete, he begins work on the interior, currently putting in cedar cabin and cockpit soles. When I ask why he's using cedar for the boat's flooring, he pushes his glasses up on his nose and says with little thought, "It's the least expensive good-quality wood you can buy. It's light, durable, rot resistant. You don't necessarily have to keep a coat of paint on it. It's good for the same reasons it's good to put on the side of your house."

And on the subject of wooden boats he is clear and optimistic: "I think we're at a remarkable time in traditional wooden boat building. People talk about the golden age of wooden boats as being a hundred, a hundred fifty years ago. But I think we are at a time when, because of the availability of material and demand, which is low, we are in a position to build the finest boats ever." With neither plentiful wood nor great popular demand for the boats, that is, builders are likely to produce not lots of weak, trashy wooden boats but rather small numbers of well-made ones.

Furthermore, he continues, you can build a wooden boat for less money than a comparable custom fiberglass boat. And because a new wooden boat requires less maintenance than a fancy fiberglass boat, he says, it can be more practical.

Duane is voluble and outspoken, but on the question of wooden boat practicality, all in this boatyard seemed to join him in disagreeing with the most common, all-but-ubiquitous consensus on them, that insurmountable notion that they require too much maintenance. This is *often* true, they allow, but not because the boats are wooden. *Old* wooden boats require continual fixing and work, and *poorly constructed* ones do, too; if you ever owned a poorly constructed wooden boat that was also old, odds are likely you're now a fierce defender of fiberglass. But the fact of the matter is that *all* boats require maintenance, and a new wooden boat doesn't need any more of it than a new fiberglass boat—different maintenance, yes; *more* maintenance, no.

Duane first came to boats through reading the book *Dove,* by Robin Lee Graham, who in 1965, at age sixteen, went to sea in a 24-foot sloop and sailed solo around the world. It ignited Duane's interest in sailing, which led to this work. Duane was seventeen when his family moved to Greenwich, Connecticut. Not fitting in any way at all into the local yachting scene—not "in my thought," he says, "in my upbringing, in my expectations, in my intercommunicative skills, in my willingness to drink"—he went to the library instead of the club. There he found the boating section and a writer named Howard Chapelle. Most regard Chapelle as a knowledgeable historian with an unbearably convoluted, at times even impenetrable prose style, but his sense and sensibilities spoke to the teenaged Duane. Chapelle wrote widely on boats from the previous century through the 1930s, the era when working sailboats mostly vanished.

"One of the things that Chapelle said on a number of occasions," Duane recalls, "was that these boats, with a little thoughtful modification, would make excellent low-cost yachts for the modern-day yachtsman who wasn't determined to look like part of the fashionable set. And I thought that was really interesting. And that's when I started thinking about doing it. I thought, There are options here. They don't have to be these glitzy racing things that you see out on the sound with their white-jacket, twelve-crew contingent."

He moved to Maine as a young adult and found a job at a yard that put out 42-foot lobsterboats for a hundred grand apiece. The builders would climb into the 1-inch-thick hulls, nothing but big bathtubs, and assemble them on the inside. Because Duane worked mainly on the wooden interiors, he enjoyed the work. He's happy now to have found G&B—"Ten years ago," he says from the cockpit of the unfinished Bella, "I couldn't hold a hammer. Now I can build a boat." And he notes that any of the men here could earn significantly more money working elsewhere, but money isn't the point. "There are many ways to solve the puzzle," he says. "It all depends on your preferences."

And there are many kinds of boats for many different preferences. While Duane believes that wooden boats are good and practical, he doesn't claim they're superior to other kinds. Indeed, he'll bend over backward to argue that fiberglass is ultimately more practical and easier to fix. Put a hole in a plastic boat, he says, and all you need is a grinder and some fiberglass cloth. Of aluminum masts on fiberglass boats, he asks, "What's more natural than aluminum?" (He'd never dream of having an aluminum mast himself, though—because of the annoying, ringing *whap whap whap* of the halyards smacking against it all night long.) He's skeptical about wooden boat loyalists' holier-than-thou mentality, the attitude that annoys so many. So why does he spend his life building these boats for G&B, and now on the weekends building one of his own, and then returning every single night to sleep on one?

Duane appears to be at a loss and says only, "Wooden boats are so much a part of who I am."

✳

As Duane continues work on *Isabella,* Bob, Bruce, Jim, Ted, and Ross work on the powerboat; Robert Bennett and another G&B regular, Chris Mullen (also a veteran Alaskan fisherman), restore *Liberty,* the 40-foot sloop built by G&B in 1986; and Nat roams the woods of Cumberland Island, Georgia. *Elisa Lee* is now upside down and fastened to the shop itself. The crew must begin to fair the frames and the rabbet, put the ribbands across the sawn frames, and bend in the oak frames. The sawn frames that are now bolted on cannot be wavy, ripply, or bumpy along their bevels. All use hand planes and, because the wood is hard as concrete, must sharpen them frequently.

"Around here," Jim says, "it's never a chisel. It's always a very, very sharp chisel. *Very, very sharp.* 'Get me a *very sharp* chisel.' 'You can do that with a *very sharp* block plane.' "

And for this purpose there is an electric sharpening stone that spins like a potter's wheel on the bench next to the vertical grinding stone. A tiny copper tube opens just above it; before using the stone, you turn a handle to allow a thin stream of water to flow onto it. It's so cold these days in the shop, though, that the stone has frozen solid. Bob is kicking himself because he meant to get some nontoxic antifreeze to put in there to prevent just this situation, but the stone is rock-hard now, and so they must plane with blades that they will allow to become less than very sharp until it warms up a little.

Ross says, "I wish that thing would thaw. I don't sharpen my tools as often when I have to do it by hand." He's too impatient.

The last week of February continues cold. On Tuesday morning Ted, Bruce, and Jim stand around the stove shivering. When Ross doesn't show, Jim says, "I guess we should get to work." The radio in Ginny's office announces the temperature: 12 degrees. Jim and Bruce leave for the outer shop and climb onto the frames of *Elisa* to begin fairing the rabbet, making it perfectly smooth, especially at the joints where the keel pieces are scarfed together, a perfect but shifting right angle along its 30-odd feet on either side.

Ross strides into the shop, saying, "Sorry I'm late. These are bankers' hours—what is it? Nine o'clock?" To Bruce up on the keel he says, "How are you doing?"

"Cold," says Bruce.

" 'Cold'?" Ross says. "This is warm compared to yesterday. It's getting warmer every minute. Look at this sunlight coming through the plastic." He points, grinning, to the ceiling above Duane and the Bella. "It's almost *springlike*."

Ross is in a good mood in part because he and his son, Lyle, are heading up to Maine tomorrow, Wednesday, to visit Kirsten for the rest of the week. But he's also charged by the bright morning and the full day of work ahead. The crew stops to listen as he addresses the day's tasks: "Let's fair up this rabbet, fair up the frames." He walks along the port side of the frames and stops at station five, the one that Bob and Bruce had the most trouble figuring out because the bevel there was so slight. "I don't understand that bevel," he says. "The boat's getting bigger here, but it's getting smaller here." He stares at it some more, then continues, "Well, fair everything up, just the first three feet of frames, so we can put the ribbands on them."

Along the first 3 feet, from the keel out, spruce ribbands will be screwed into the frames. These 2-by-1-inch lengths of soft, light, flexible wood run fore and aft across the frames and will serve as a mold for the bent frames. Ribbands are also used to check the fairness of the frames. Two people typically work in tandem to fair frames, planing them and then holding a ribband across three frames, mimicking a plank; wherever the ribband doesn't rest flush, wood must be planed off. They hold the ribband against the frames, squint, take the ribband off, eye it some more, and plane some more.

Bruce asks, "We're going to have two steamed frames between stations?"

"Three," Ross answers. "But we're only going to put one floor timber on." Ross steps back and looks at the whole boat. He's thinking about weight. This structure is going to get heavier and heavier, starting right away, when he puts the transom on, and then again as the ribbands are fastened, and steamed frames are bent into place, and floor timbers are bolted, and finally as the planks are hung. "Are we satisfied that this keel is supported well enough?" he asks. His question doesn't sound rhetorical, but practically speaking, it is. "It's going to weigh three times this much. More: four or five times. Ten thousand pounds when it's planked up." He instructs the crew to get posts under the keel fore and aft and toe-nail them in—final supporting pieces. He wants the plugs made to fill the holes in the keel. "Use teak," he says. "Teak's every bit as durable, and it won't dull the blade as much as angelique." He tells Ted and Myles—Myles Thurlow, the high school student who's apprenticing here as part of his studies, recently back from seven weeks' sailing in the Caribbean—to restack wood in the parking lot, separating out the wood Duane has bought. "Then we'll get some long planks, and you guys can rip them into ribbands."

And the day is begun.

The outer shop is bright; all are industrious after Ross arrives. Ginny is snug in her office; it's tiny, so a space heater works efficiently. She claims to feel guilty, gazing from her cozy perch at the boys working with frozen hand tools, their breath all but crystallizing in mid-air and falling to the lofting floor with a shimmer. Her radio hangs against the wall and is tuned to a classical station—Mozart and Haydn all morning long in a toasty office.

When Ginny does step out, in late morning, I am at the drill press cutting plugs for the keel. I have found some dark scraps of wood, their surface dull, almost waxy, from which to cut the 1-inch-wide plugs, and as I lower the cutter into the wood, aromatic smoke begins to rise. As Ginny passes, she says, "You can tell what kind of wood that is just by smelling it. That's teak—smells like olive oil, a little bit."

✳

By late afternoon Bob, Bruce, Jim, and Ross begin concerted work on fairing the frames. The purpose of fairing is twofold: to reduce surface area and therefore the potential for deterioration, and to ready the frames for the planks, which must rest flush against each frame as they bend along the hull from stern to bow. Fairing is long, slow work; the skill and eye for it come only with practice. From one angle a frame can look perfect; from another, ripples and dips are apparent. Ultimately each frame must be close to perfect, not only in itself but also in relation to the frames on either side of it.

"It's the hardest thing to teach people," Bob says. "You've got to step away and look at it from a distance."

Ross agrees: "You can't do a good enough job fairing. It's a job you can keep doing forever, because it's impossible to get perfect."

When Bruce begins fairing his first frame, he says, "This is going to be hard. These grains are going in opposite directions"—a circumstance he didn't foresee when he was bolting separate pieces of angelique together to make these frames.

Ross says, "Yeah, plane across. You don't want them splintering up. Be careful going at it. Look at the big picture, not just what you're working on, because if you're not careful, we can wind up with a big mess. And this thing has been setting up so nicely, I don't want to do that."

All day long, Ross will repeatedly stop working and walk back and forth to survey the others' progress, once calling Bruce and Jim's portside frames "wavy." And then he makes a guess: "We should be able to roll this thing over in six weeks"—mid-April. "The better we do this job now," he adds, "the faster the rest of the work will go. We won't be stopping to pull out a ribband to fair a frame."

Late in the day, when the sky darkens, overhead lights are switched on. A deep chill returns to the outer shop, and breath appears bright and smoky in the electric illumination. The appearance of Brad Ives gives everyone an excuse to stop fairing and gam awhile. He's just back from Suriname and has stopped by to pick up a check for the wood that's been delivered. Ives strolls laconically across the lofting floor. Duane steps off the Bella—he owes Brad money as well.

"How much do you need?" Duane asks.

"Everything you got," says Ives, dead serious from the look of him. Then he smiles slightly and says, "If you can pay me for the deadwood, three hundred seventy-five, that'll be fine." Duane has designed his own 38-foot motorsailer and will start to set it up in the spring. The deadwood is a big chunk of wood bolted directly behind the lead ballast on a sailboat. The balance of Duane's order will arrive soon in Mystic, Connecticut, at the seaport and museum there. Mystic, currently building a replica of the slave ship *Amistad,* is another of Brad's clients.

When Duane hears this, he says, "That's the worst place for that wood to go. They'll hide it!"

Brad laughs softly and says, "I'll be there to meet it."

Whenever Brad returns, the topic of discussion is always Suriname and its wood and what Brad has done and seen. Each one of these boatwrights is a traveler and delights in stories from distant lands. Ross wants to know where this new delivery comes from, how Brad found it, what part of the tree's been used. Brad talks mainly about curves and sweeps. For these he traveled three hours over deeply rutted rain-forest roads to the Amerindian villages, to explain to those indigenous people exactly what he needed. The Amerindians used to walk through the forest with chain saws in hand, cut down trees, take their stumps for Brad, and leave the rest to rot there. Brad has had to convince them that this is an unnecessary waste and that they need to bring the entire tree out of the forest. The Amerindians have no heavy machinery, so they must drag what they cut out of the forest on foot, over dense, uneven, often swampy terrain. Thus they can carry logs only up to about 8 feet long; beyond that, the trees are too heavy. Brad also talks about angelique, called basra locus in Suriname. These trees grow to about 150 feet, with a 70-foot bowl at the rain-forest canopy. He explains that about midway up their trunk, angeliques curve briefly, then continue straight up. He tries to find these curves, but because such timber is not in high demand at the mills, it's difficult to find, and it's also difficult to ship because this shape is inefficient in terms of filling a rectangular container: managers of the shipping concerns want long, straight lengths, and therefore more board feet, to fill their containers, not those inefficient bends and curves.

Duane hunts through his bag, which is set on a plank in the corner of his work shelter, locates his checkbook, and writes out the amount. Discussion ebbs; the air grows colder, and Ross and the others return to work. Brad leaves at a stroll that couldn't be slower if it were a summer evening on the beach and he had nowhere else to be.

✴

"Back in the days when we actually sailed occasionally, we planned to do a couple months bopping around the coast of Maine," Duane says one evening after work, staring out at the Coastwise dock and beyond to an arriving ferry. I've asked him for a sailing story, and he's chosen a trip he and Myrtle made in their 36-foot L. Francis Herreshoff ketch. "We left early in the morning from Vineyard Haven. It was one of those glassy, flat-calm mornings, dead, but by midafternoon there was a pretty good breeze—twelve to fifteen knots, all sails up, easy sailing. None of this hanging from your shrouds with your teeth while trying to reef the unruly main! Just the sound of the water lapping against the hull. Even after the sun set we had the first half of the moon, a perfectly clear sky, reflection of moonshine on the water. Sailing on through the early part of the night, and you could see the Provincetown Light blinking away in the distance and slowly moving astern, and as it goes astern, we're out into no land visible—far enough where the curve of the earth makes all the land disappear. A perfect evening sail.

"But as we got up toward the Gulf of Maine, the fog started coming in. We could see it begin in the covering of the moon and the sky, and for a few minutes we thought, Should we return to Provincetown?"

Sailing in fog can be more dangerous than sailing in bad weather and is more nerve-racking. Great, crashing waves provide their own tension release: you're actively sailing, or actively hove to and hanging on, and you know where the danger lies. Storms aren't easy, and they can be terrifying, but fog is something different. The danger is in its stillness, in its blinding you: small boats, too tiny to register on a freighter's radar, can be run down, or you can collide with a floating object or a landmass. But fog also turns ordinary sounds into signals that can seem at times supernatural. It's impossible to gauge how far off that foghorn is, or that ringing buoy, or the water pounding against those jagged cliffs; it's difficult even to discern which direction a sound is coming from.

"We're dead reckoning," Duane says. "The last time we had a fix was when we left the Cape Cod Canal. The fog is coming down, and it's getting darker because the moon is going down. Fog descending, and everything starts to get damp. A little after midnight the wind starts to die, so, not being too proud, we decide to start up the engine, just chug along slowly through the night. By now we're in a cocoon. The fog isn't superthick because you can see from one end of the boat to the other. We weren't really concerned about running into anything because it would only be another boat and they'd have their lights on. We did this for the rest of the night.

"The sun comes up, but you don't see the sun because you're in this fog, waters rippling away. Breakfast time comes, and again we're sitting on deck. We're not talking much because of the noise of the engine.

"Now it's getting to be ten o'clock, and from my dead reckoning I'm saying we're within a few miles of Portland. Tension was starting to build up, because as you approach Portland, the land is starting to close in on you from either side, this big scoop of land: the southern part of Maine is curving in from the west, and whereas it has been sandy up till now, it's starting to get rocky. Portland is the beginning of what you think of as classic Maine coast. It's also a busy harbor, not only for small boats but for tankers. Anybody going in or out of Portland is converging on this same spot. We're now back with engine off, ghosting along. But if my dead reckoning is right, we're only four or five miles from the big buoy at the entrance to Portland, and all around us we hear different kinds of sounds.

"There's a *beep* here and *honk honk* over there. There's stuff going on. We can't see any of it. Sound travels tremendously in fog—it can be very deceiving. I've been in places where a lighthouse has been several miles away, and a thick fog rolls in, and it sounds like it's right there, though you know very well it's six miles down the coast. Once in a while we hear a motor, we hear an AM radio blaring out. We listen to the news for a few minutes and a couple of dumb songs and then that drifts away—some fisherman tending his traps."

And then a voice came over their VHF radio, Duane remembers. The voice sounded Norwegian or Swedish as it announced the name of the vessel, a 500-foot tanker, and its specific course into Portland. This was interesting, Duane says, because according to his chart, and his dead reckoning, the course that the tanker captain had named should be right about where *they* were. In fact, maybe that was what those horns were that they kept hearing. Duane and Myrtle now became hyperalert in the enveloping fog, listening and watching for anything, but especially a 500-foot iron box plowing west. A tanker wouldn't necessarily see them if they were in its way, and even if the crew did spot them, they wouldn't be able to alter course fast enough to avoid destroying them. Nor would Duane and Myrtle have enough speed, either by sail or by motor, to get out of the ship's way once they saw it. They were blind, did not know where they were or if they were on course or how long this might go on. It felt as if they were under some kind of spell, in a disturbing dream. The eerie noises continued, and they remained perfectly motionless, waiting, all senses pricked up.

Ten yards away, a whale surfaced. They didn't hear a thing; suddenly the form was just there, and it paused so its blowhole was just out of the water. The whale breathed loudly, the air rushing in, then rushing out again. They were close enough to see the blowhole expand with the rush of air, in and out for a few breaths. And then the whale rolled back down into the sea, its entire length curving through the air beside them. Duane and Myrtle were transfixed. The whale's body kept going and going. It had to be a hundred feet of dark, shining, curving whale back. It seemed endless to them. Duane had never seen a whale so close, never shared the same spot of sea with one. It was an awesome sight, and for minutes neither Duane nor Myrtle thought of anything—not the fog, not the tanker, not landfall, not where they were—but this magnificent creature that had risen to save them.

"After the whale was gone," Duane says, "the fog lifted. Just *went*. It didn't go up very far; the visibility maybe went up to between one and two miles. But two miles—there was our buoy!"

No tanker was in sight. Myrtle had been doing some serious prayer work on the subject of that tanker. Had it been an hour since they'd heard the voice on the VHF? Duane couldn't say, but surely it had been long enough for the big ship to make it into Portland's outer harbor as he and Myrtle drifted under sail in light winds.

"We practically hadn't gotten the anchor down when the fog came down thicker than ever. You could hardly see, and you really didn't want to be out there in that kind of fog. It didn't lift for two days."

VII

Ross returns from Maine as scheduled and is back at work on Monday morning. The winter storm predicted for Thursday night arrived hours early and dumped so much snow on the island that only those who could walk to the shop from home could work, and with both bosses gone, few had any motivation to do that. On Monday Bob says, "I'm glad Ross is back." This is because work now resumes in earnest, with redoubled effort for the lost days. Ross gives morning instructions to cut extra floor timbers, one for each bent frame at stations where the engine will rest and where the mast will be stepped—"places that will have a lot of stress," he says, adding, "Weight is a consideration in a powerboat."

Bob says, "Think it will go ten? I think it'll go ten." A speed of 10 knots, or about 11 miles per hour, should be sufficient in the waters of St. Croix.

Bolts between 14 and 24 inches long, two for each timber, must be fashioned; Ross determines their measurements from the drawing on the lofting floor. Jim begins to plane thick pieces of angelique for the floor timbers. The tropical wood has a distinctive odor when it's freshly cut or planed. Jim puts his nose close to the wood, winces, and says, "Zees angelique smells like where za cows are on za farm."

By Tuesday morning Nat has returned to the shop as well. He enters wearing a green vest, green wool sweater, green beret, and pale-orange Carhartt work pants, and carrying his canvas briefcase. Jim is running a plane across the edge of a floor timber he's fixed in a vise,

Bob and Bruce are discussing the timber Jim is planing, and Ted and I are listening. When we see Nat, we all say hello and "Welcome back."

Nat says, "Just like when I left. Four people watching one work." He chuckles and says, "I hear you got a little dusting." Leaning toward Jim he says, *"Petite neige?"* The island is still digging out from under it, work crews moving trees felled by the weight of wet snow.

Nat heads toward the outer shop. "I see you got her turned upside down." He regards *Elisa Lee,* then has a look at the Bella. Duane says, "Welcome back!" "Thanks," Nat says, and then, "This is looking good." He and Duane discuss the toerails and the cabin trunk until Ginny scoots past them and turns left into the shop, headed for her office. As she passes, Nat says, "Well. *Ahem!* I better see what Mrs. Jones has for me."

She doesn't break stride, only shouts, *"Plenty,"* and is gone.

Nat's brow lifts, and he says, "Ah-hah," and follows her to the office.

Ross is already there, and when he sees Nat he shouts "Hey!" with genuine gladness.

"Good to see *you,"* Nat responds.

"How was the trip?"

"Great," he says. "A lot of fun." He and Pam covered many miles on foot through the Georgia woods, he says, and saw all kinds of animals, including wild pigs and armadillos. Then he says, "Oh, those woods! Longleaf yellow pine, hickory, live oak. It's a boatbuilder's *dream*—to see every shape of a boat in those trees."

When Nat hikes through a forest, he sees boats.

✻

In addition to the two new constructions, a major repair has been under way all winter on the 40-foot sloop *Liberty,* which Nat designed and G&B built in 1986 for Primo Lombardi, the owner of a pizza joint in Oak Bluffs. Likely the first American boat to be built of Suriname timbers, she's a sweet vessel, as just about all Nat's designs are—simple, powerful, with moderate displacement but also relatively low-riding and thus sleek-looking and surprisingly quick and easy to maneuver.

WoodenBoat magazine has critically reviewed only one of Nat's boats, a 30-foot yawl named *Candle in the Wind,* but the writer of that 1992 commentary was Joel White, one of the most respected designers in the industry and the son of the great literary stylist E. B., a man who not only devoted his life to wooden boats but also looked at boats with an eye toward what might be lasting about them, what truths they might point us toward. In Nat's design he apparently found abundance.

White began his review by calling G&B a boatyard "of the old school" and describing the new yawl as a "fine example of the quality of design and craftsmanship that has made Gannon & Benjamin well known to a small but discerning circle of admirers." White had contacted Nat for some background on the boat, and Nat, addressing the design requirements given him by the Englishman who commissioned her, had written in a letter, "I am convinced that performance, comfort and looks are all compatible, and to design and build a boat for a family to enjoy is the most reasonable request."

White looked carefully at the attributes of design and the aesthetics of "this old-fashioned boat that I believe has implications for the future." One of his convictions, and a perpetual theme in his designing and writing until his death in 1998, at the age of sixty-six, was this: "The greatest fun in boating usually comes in the simplest boats."

"Most modern boats," he continued, "are simply too complicated. They are so full of systems, which too often fail to work, that the feeling of self-reliance—that wonderful ingredient in the pleasures of cruising—is now missing. The modern cruising boat makes the owner a slave to the systems and to the chore of keeping them all working. The Loran isn't working?—well, we can't sail without that. Call the electronics man." White's command: "Keep it simple."

Candle was one such example of old-style simplicity. "The hull reminds me a bit of small English cruising designs of the 1930's," White noted, and he went on to remark on the design elements—the things that made the boat tick, as Nat would put it—including the hollow from the garboard up to the firm turn of the bilge, the displacement, the waterline length, the displacement/length ratio, the slight cutaway of the forefoot, the rake of the sternpost, the fact that the 4 feet 9 inches of draft allowed the 2-ton lead-ballast keel to be low enough for good stability. There was "nothing revolutionary in the lines plan," he conceded, "but a very nice combination of elements to produce a fast, shapely hull that looks right under its old-fashioned gaff-yawl rig."

The proportions of that rig were, White wrote, "perfect." "So many new boats built today with gaff rigs suffer from ill-proportioned spars, clumsy rigging details, and a lack of knowledge of how things were done a century ago. *Candle in the Wind,* with her eye-spliced shrouds and wooden blocks, would not have looked out of place in a turn-of-the-century regatta."

White concluded with the same simplicity that he espoused: "I think this little boat is very close to perfect."

Five years later the man who commissioned *Candle,* H. Nicholas Verey, died of leukemia, and family affairs had to be settled, one of which was selling the boat. Happily for Vineyard Haven Harbor, she was bought by residents of Martha's Vineyard and returned to live here year-round under G&B's patient care.

White might have made similar comments regarding the simplicity, lines, and design of *Liberty* or most any of Nat's boats—they all shared these attributes. Benjamin-designed boats weren't all perfect, and some were quite a bit better than others, and even within the same design—the Canvasbacks, the Bellas—some boats were better than one or another sister. Ross admired Nat's designs but was not without his opinions about them. The sloop *Swallows and Amazons Forever,* with her slack bilge, never really worked for him—he thought her homely, felt sorry for her. Nat himself was not particularly fond of his cat-boat, *Seasons.* But then he didn't like catboats generally, so perhaps it had been difficult for him to muster the passion for that distinctive style and rig.

Liberty, though, she was a fine vessel, tropical hardwoods bent into classical lines. And now she had a new owner, Doug Cabral, the editor of the *Martha's Vineyard Times,* who'd purchased her with his wife, Molly, and another couple. The previous owner, according to word around the yard, had wanted just a little more boat than he could actually afford (he was boatstruck), and he hadn't maintained her well or even sailed her all that much. Cabral and friends had sailed her for the summer with the agreement that G&B would haul her out for the winter and restore her to perfection, or, in the words of Cabral him-

self, make her "a proper boat finally." So there she stood, on blocks beside the shop, towering above all as they took their ten-thirty coffee
break. A portable staircase rested against planks, staging that had been
built around her waterline. To get to her, you took fourteen bouncing
steps from the dirt up to the staging and from there ducked beneath the
red, white, and blue tarp that sheltered Robert Bennett and Chris
Mullen, and soon Nat, in their work.

✳

Doug Cabral, seated at his computer in the second-floor offices of
the *Times,* has been able to turn to the left and look out a sliding porch
door and watch his boat being worked on all winter. He admits it's a
distraction, but not half as big a one as the boat will be this summer,
when he'll see her moored just off the G&B dock, tipping back and
forth in the harbor, nodding to him at his office desk, beckoning like a
mistress, "Come to me."

Cabral, a tall, beefy fifty-three-year-old with short gray hair and
the skeptical, no-nonsense demeanor of any self-respecting newspaperman, runs a good paper; both he and the *Times* are well liked on
the island. It's one of two papers published here; as a glance at the clas-
sifieds indicates, his is the year-round newspaper for the island resident,
whereas the *Vineyard Gazette,* a broadsheet, caters to the summer people. Both are decent small-town newspapers, though, and Cabral is
perfectly suited to running the news for the year-round Vineyarder.
He's been here since 1969, when he delivered boats and free-lanced for
boating magazines. Thirty years a resident, with a wife and children, a
dog, and a house up-island, he's seen the changes in the Vineyard, its
transformation from a secluded rural retreat to a tony, wealthy, over-
crowded destination, a destination that he argues almost always disap-
points the first-time tourist.

"Unless you live here," he says, "there aren't many great beaches,
and you can't get a drink in four of the six towns."

Oak Bluffs and Edgartown are the only towns that sell alcohol, and
the island's best beaches, such as Lucy Vincent in Chilmark, are indeed
open only to residents of specific towns and strictly guarded from June
through September. Furthermore, the island is no longer, in Cabral's

words, the "hard-bitten, antique outpost" that longtime Vineyarders like to imagine it is. Martha's Vineyard has become, according to him, suburban.

Having arrived before both Nat and Ross, Cabral has been able to watch their presence and partnership on the island grow, so he is, from his perch above the boatyard, an ideal and appreciative observer of the two men.

Nat, he says one wet March morning while *Elisa*'s frames are being faired and readied for the ribbands, "marches to a different drummer. . . . He's a mild, decent man." His personal life is steady and predictable. Ross, in contrast, is "mercurial" and "rigid." When Ross moved houses for a living, Cabral remembers, his main problem-solving method was to find a heavier sledgehammer. He senses that perhaps "Ross feels uncomfortable in a house," musing that he may be better suited to the forest. "Ross is none of the things you want in a modern human being," Cabral concludes with a grin.

He admires and respects both men and says that to those longtime residents who summer in ramshackle cottages passed down through the generations, "Nat and Ross are an important ingredient in their memory when they're away from this place." Their "unusual, anachronistic work" jibes, he says, with that antique-outpost *notion* of the island.

It is also true that Cabral loves wooden boats and is grateful to Nat and Ross for their work. "I was infected young, and I've never been able to get rid of the disease," he says. Asked why he favors wooden boats, he doesn't wax nostalgic or fly off the handle about cathedrals and curves; he simply says they're what he prefers. Some people like Picasso, some prefer Goya, he says, and it's the same thing with boats—sometimes you just know, "That's for me." The newspaperman is not boatstruck, but throughout this winter he's nevertheless been *itching* for summer, when *Liberty*—"one of Nat's cleverest designs," he says—will be back in the water. "Nat is not *distracted* by what's going on in marine design," he explains.

Cabral is furthermore a great admirer of Nat and Ross's craftsmanship and method, applauding their simplicity and praising both men for being "very conscientious in the construction stage. They're very honorable in putting a boat together. It probably reduces their profit margin. If they *have* a profit margin. It's an achievement these days to do decent and authentic things."

Cabral scratches his head at how the partnership of these very different men can be so harmonious, but somehow it works, he says: Nat the public voice, the designer and fine craftsman; Ross the brilliant mechanic, the engineer and cast-iron motor of the operation. "Nat is much better at the finish work," Cabral says. "The work Nat does will look better than the work Ross does, but Ross will get it done faster." And for *Rebecca* he reserves the highest praise: "It's a gorgeous piece of work," he says, citing the extraordinary joinery, the exquisite lines of the planking. The whole boat, he says, inside and out, seems to have been built to the standard of the fine finish work on the inside.

Cabral takes periodic strolls through the shop, trailed by Copper, his Rhodesian Ridgeback, to gaze up at *Liberty* and to enjoy the anachronism of the shop and its workers.

Even when Nat and Ross themselves are nowhere to be seen, their differences are evident in their toolboxes, which float here and there about the shop and personify exactly the same qualities that Cabral points out. Nat's toolbox is a spacious rectangle whose handle-bearing ends curve up and out in an hourglass shape; the handle itself repeats this curve, treating it as a design element—wide at the sides, narrowing at the grip—and is attached with a perfect dovetail joint. Nat composed the box out of three woods—pine, maple, and oak—whose visual contrast is now diminished by a decade's use. Most revealing is the handle, which is flat so Nat can use the box as a surface, a mini-sawhorse, when he's working in a cockpit or small interior space. The box is elegant in design and construction, but its ultimate elegance is in its function. This box is very much Nat Benjamin.

Ross's toolbox is a dull, uniform brown, splintered and ragged on the sides, pocked, chipped, scratched, scraped, and so loaded with heavy tools that Ross has had to attach with duct tape a supporting bar of wood to the broom handle running across the top to keep the thing from crashing to the floor when he lifts it. But even this beat-to-hell box contains an elegant logic. It's a working toolbox, meant to carry tools; it isn't supposed to be art or a fancy showpiece. Indeed, any time spent making such a thing fancy is, to Ross, time *not* spent getting the real work done. Now and then Ross looks at this toolbox and says, "It's really time for a new one," but then his eyebrows lift and he gets back to work. As long as it does its job, there's no real reason to build a new one. Ross is a man who cares strongly about toolboxes, who knows that the workman is, or should be, invested in his toolbox; he therefore instructs every new apprentice to build his own box as his first order of duty, and he points him to the scraps of tropical hardwoods stacked against the wall beside the wood-burning stove.

✴

On the day of Nat's return, Ross walks out to the "Pepe's" pickup and retrieves a heavy construction of metal and brings it inside to show Nat. He sets it on the table saw to have a close look at the thing. It appears to be composed of only a few working pieces, a central threaded shaft an inch or two thick, like a giant screw, with parallel arms, connected to a large bronze disk that slides up and down the length of the shaft—or is *supposed* to slide. The piece seems to be from the era of the metal lathe and planer, the era of cast-iron machinery, and it's fairly well coated with either dry, oily sludge or rust, except where there's the corroded verdigris of the oxidized bronze. It's the mechanism by which a rudder is turned from the wheel in the cockpit by means of a worm gear.

"I'm giving this to Doug," Ross says. Cabral wants to remove *Liberty*'s tiller and replace it with a wheel. A while back Doug gave Ross a table saw he wasn't using, and Ross considers this worm gear a return gesture.

Several years ago Ross read a classified advertisement for a boat in Fort Lauderdale. Wooden boats, if you don't take care of them, will sink, and this one had sunk right in the marina where it was docked. The ad was in *WoodenBoat,* placed there by the owner, who hoped to cut his losses. The boat was free for the taking to anyone willing to haul it out, the ad said. Ross paid four grand to have it transported to Vineyard Haven, wanting its parts, including the lead ballast, which he's using for his own boat. This worm gear also came off that boat. It lies now on the ship's saw as if stilled in a sepia photograph. Certainly it's nothing that could work today.

But Ross grabs a can of lubricant, shakes it, and begins to spray the metal pieces. Nat gets in there, too, scraping away built-up crud. Ross turns it over. Nat hits it with more spray. Gunk slowly comes off; lubricating oil pools beneath it. Slowly the bronze disk and threaded shaft along which the disk moves budge, coming back to life. It's as if the two men are reviving a living organ. Soon Nat has the whole thing turning smoothly and without resistance. He tries to jiggle the disk that rides up and down the heavy thread; there's no play in it. "The thread is very good," he says. "It's very nice."

"These simple mechanics," Ross tells me, "are not prone to malfunction. Now steering on boats is more like it is in cars, and it's very prone to malfunction."

This is why Nat and Ross regard the old worm gear as such a special piece: only a few moving parts made of steel and bronze. You could depend on it almost forever. "We took this off a boat that was built fifty years ago," Ross says, "and are putting it in a boat built fifteen years ago. And we'll expect it to outlast us."

This is the perfect summary of the way their world works, the only way they will have it work. The way they determine it will be. Their world has few moving parts, and what parts it does have are meant to last. It is true of the worm gear, all but breathing once again in a puddle of fluid on the ship's saw; it is true of their toolboxes, and the Bella and Elisa Lee in the outer shop this very moment, moving piece by piece toward their summer launch dates; and it is true of the halted Rebecca. The worm gear represents the reason they don't rely on fancy equipment, modern steering devices, electronic navigation systems, or anything made of plastic. Such things are unreliable, and if something is unreliable, it is no good—period—even when it's working. If you can't depend on something absolutely, then it is by definition temporary, and something temporary is something that needs to be fixed or replaced. These builders use only tools and equipment and build only boats that can be depended on absolutely. Doing otherwise is, for them, unnecessary.

Recently David McCullough asked Nat for help with some pages he was working on. McCullough is the author of Truman, the best-selling, Pulitzer Prize–winning biography and paean to the thirty-third president of the United States. Millions know McCullough's forthright voice from Ken Burns's documentary *The Civil War*, and still more know him as the host of *The American Experience* series on PBS stations across the country. That is, if they haven't already read any of his fat, compulsively readable books about the building of the Brooklyn Bridge and the Panama Canal, and other historical events. He was currently at work on a book about America's second president, John Adams, Adams's wife, Abigail, and their times. When he arrived at a particular event in Adams's life, he called Nat for some insight into navigation in the eighteenth century.

Still standing with them at the ship's saw and the worm gear when this information comes up, I ask how either of them could advise one of the country's best and most popular historians on techniques of eighteenth-century navigation. Nat and Ross reply in happy unison, "Because it's how we do it now!"

Nat says that he and Ross both chartered and delivered boats in the 1960s and 1970s, before the advent of the Global Positioning System, when the most complex navigating device on a boat was likely to be a Loran or radio direction finder, both of which were "pretty much useless," according to Nat. Nat had taught himself how to navigate by sextant in his early twenties, and Ross had, too, so they never had to rely on electronics to get safely to and from distant points of land, so long as they had that ancient navigating device, which relies not on a manmade satellite launched into space and wires and electricity on a boat, but rather on the seaman's own eyesight and the position of the sun (which latter Nat and Ross consider to be a pretty permanent and dependable measuring device). The stars, too, can be handy jigs for making your way across the surface of the earth.

But Nat did something even more valuable for McCullough than confirm details of navigation, refine descriptions of the types of waves a captain would encounter on that part of the ocean (McCullough remembers that Nat asked him to reconsider describing the sea as "turbulent," for instance; the sea in question is now "steep"), or explain why the North Atlantic was so treacherous in winter, the time of the voyage in question (because there were no sophisticated weather-detection systems then, ships were often blindsided by monster storms: "If you're out there," Nat says, "and you got hit by a nor'easter like we had last week, and you're above Cape Cod, you'll be driven up on the rocks, and everyone will die. There was no emergency rescue. It was dangerous").

In the end, Nat's most significant help involved some historical reconstruction.

The time was February 1778, in the middle of the Revolutionary War, and Adams had to cross the ocean to France to help Ben Franklin with treaty negotiations. He brought with him his ten-year-old son, John Quincy. The dead of winter was a time when few ships went to sea. Moreover, the British navy was blockading the colonies precisely to prevent any attempted passages. Adams might easily have been captured, in which case he'd have been delivered to the Tower of London and likely hanged. The city of Boston teemed with Tories and British spies, so Adams could not board the ship there; instead, the ship was loaded in Boston and then sailed off the black, rocky beaches of what is now Quincy, where Adams and his son waited at dusk in a driving snowstorm.

The voyage proved to be every bit as difficult as might be expected given the winter and the war, and then some. Just about everything that could go wrong on a voyage did. Working from the log of Captain Samuel Tucker and from Adams's own diary, McCullough's narrative describes a nightmare crossing. The ship, commissioned the previous year by the Continental Congress, was a U.S. frigate, a fast war vessel more than 100 feet long, carrying sixteen guns. On this voyage, it was caught in a tremendous storm; it was hit by lightning; the main mast split in two; crew were killed; and a battle was fought with a British ship.

"I wanted to get it right," McCullough tells me. "And I wanted to make it as horrible and vivid and exciting as it really was. So I asked Nat to read it for me. And I think this is a measure of the kind of guy Nat is, the way he approaches everything. He got up a chart of the North Atlantic, and by using the captain's log, he plotted the whole voyage of that ship, which I'm sure has never been plotted before. He determined for me, as I now say, that the storm actually blew them two hundred miles off course. That's a lot of miles off course." This voyage, which has been written about in only a cursory way until now, was an important event in the history of America and a fundamental experience in the life of one of the country's founding fathers. Writing to Thomas Jefferson many years later, Adams would describe the crossing as a metaphor for his whole life. McCullough is the first to write in detail about this significant journey, and he is indebted to Nat for his counsel and expertise: "He was extremely helpful, but he was also clearly enjoying all of it, and that's what he conveys in his work. He's enjoying *all* of it. He loves it. You can tell a lot about somebody by what they love, and he loves that work."

✳

Nat and Ross built a Bella for McCullough, named *Rosalee* for his wife. McCullough loves that little boat so much that he and Rosalee sometimes row out and just sit on her, don't even sail—she's simply a lovely object to behold and a lovely object to behold things *from*. The boat, he says, is a joy to own: "I would like to own that boat even if I never went out in it, just to look at it. The way one would love to own a painting or a great piece of furniture."

McCullough has known Nat for twenty years and Ross for longer, and so I paid a visit to his home in West Tisbury—classic New England architecture with eighteenth-century origins—to hear this prominent historian's thoughts on the Bearded Ones. McCullough is not quite six feet tall, but he seems to tower in the door frame because of his snowy, precise hair, his bright-blue eyes, his hearty complexion and handsome visage, and because of that gift of a voice, a voice that's resounding without being loud, a voice that carries a natural conviction and authority in its cadences. He wears a crisp blue oxford shirt and a navy cardigan sweater vest, buttoned, exactly what you'd expect of a past president of the American Society of Historians.

The first room he shows me is the kitchen in back, explaining that it may be the last four-cornered construction completed by Ross Gannon. The antique house has an almost Quaker spareness to it, but it's filled with elegant furniture, and the walls are hung with paintings, many by McCullough himself.

We sit in his living room, whose walls are lined with books, a watercolor of *Rosalee,* under construction at G&B, leaning up from the floor against volumes of history. In his work, as in his conversation, McCullough concentrates on the amazing and the remarkable. He ele-

vates lives and events, holding them up for the reader at their most
thrilling angle so that the light catches them just so. He prefers to focus
his considerable skills as an observer of history, as a writer, on what is
best, as opposed to the good and the bad in equal measure, or simply
the trashy and sensational.

And so he is delighted to talk with me about Nat and Ross, saying
straight off the bat, "I think they'll be writing about Nat and Ross a
hundred years from now. They are truly creative geniuses in their
field."

McCullough doesn't want to romanticize Nat and Ross, he says, or
paint them as Martha's Vineyard hippies trying to be alternative. "I
know nothing about their hippie days, their seedier youth." He chuckles. "I have no doubt they weren't exactly Boy Scouts.

"I think Nat and Ross are the way they are," he continues, "because of what they do. We're shaped by how we go about earning
our daily bread. We *become* what we do. We are shaped by our choice
of vocation and what demands that puts upon us, what expectations,
standards—ethical, professional.

"They're reminding us how much we're losing in this homogenized, marketing-ethic, throwaway culture we're in. They're asking the
owner to pay more money and pay more attention to what he's doing.

"They're *incorruptible,*" McCullough says. "They really are. That
comes from working with their hands and loving what they do and
dealing with the elements. They're aware of which way the wind is
blowing. The *real* wind, not just metaphorically. The real wind. The
tides. They're like a farmer or a hunter: people who know who's *really*
in charge.

"Nat and Ross are like people were a hundred years ago. We forget
how *tough* the elements were. We forget how tough *people* were." Mc-
Cullough remembers encountering Ross in the shop and seeing that
he had a badly wounded hand—in fact, a burn that had become infected. "Ross, what happened?!" he asked. Ross acted as if it were literally a scratch; he hadn't had it looked at by a doctor or even bothered
to wrap some gauze around it.

For McCullough it's a matter not simply of toughness in a historical way, but rather of toughness combined with a knowledge about
how the physical world works. The more he learns about eighteenth-

century country people like the Adamses, the more he admires how much people knew how to do, just as a matter of course. Nat and Ross are like that, he suggests.

"They are men of genuine integrity," McCullough says. "They are trying to make what they make as well as it can possibly be made, without any pretense, without any artifice, without violating the soul of what a boat should be in wind and water.

"Also, because they're so good at what they do, they have a confidence that gives them an exceptional degree of calmness. They're very calm, even in the face of things' going wrong. They're not show-offs, they're not baloney artists. They know who they are and what they do. They're *grounded,* as Quakers sometimes say; they have an inner ballast that comes from their work. They're inner-directed, and such people are rare in our time. They are *trustworthy.* I would trust Nat Benjamin with . . . anything. I'd certainly trust him with my life. I would trust those men—if you were going to do something really difficult, if you had to go across the country in a covered wagon and face unexpected dangers, hardships, the need for innovation and resourcefulness, could you pick anybody better to have with you than those two guys?"

McCullough summarizes his feelings, his judgment, of Nat and Ross by turning to another literary artist, the playwright David Mamet. "Did you ever see *American Buffalo*? There's a line in there: 'It all comes down to whether or not you know what the fuck you're talking about.' And they do."

✳

"Come on, I want to get this timber up," Bob says.

Not only is Bruce lax, he also loves to bait Bob with deadpan questions concerning the quality of Bob's work. "Is it perfect?" he asks.

Bob snorts and says, "It's *close.*"

"Perfect takes ninety percent longer," Bruce counters.

Jim, who is up fairing frames, says, "Zer is no such thing as perfect."

Nat walks through the shop, stops at *Elisa,* and says, "Everything fairing out all right?" Bob affirms that it is. People are continually noting how fat this boat is; most are used to seeing a slim sailboat hull filling the shop. When Nat, the man who made her fat, hears this, he says, "It's gotta have that. If it's too narrow, it'll just dig a big hole in the water." He clears his throat and heads up the portable stairway and into *Liberty*.

Bob says, "As soon as we get this faired, Ross is going to line it off."

Lining off a boat is the first stage of planking. Ross must determine how many planks there will be and how wide each one will be at any given station—in effect, draw the shape of each plank onto the skeletal hull. The planks will each have their own individual shape; each will have an edge beveled to make room for the caulking, and many—those fastened at the turn of the bilge—will be hollowed, or backed, out. Ross will also mark on each station the important lines, some of which won't correspond to a single plank, such as the load waterline (where the boat will actually sit in the water), and the sheer as well as the sheer at the raised deck. These marks will also determine the ribbands, the long strips of spruce that will be put on temporarily for the purpose of bending in the oak frames.

The first thing Ross will do is divide the hull into three sections of more or less equal length: the bottom of the boat, the turn of the bilge, and the topsides. Much of the bottom and top will be flat and can therefore take wider planks. The turn of the bilge includes a severe curve at the after stations and therefore requires narrower planks to reduce the amount of backing-out that's required.

Once he's marked those dividing lines, he will then consider station five, where the girth is widest, decide the maximum width of any plank at that station in each of the three sections he's marked, and simply divide the length by that number to come up with the number of planks for each of the three sections. He will then divide the length of each station by that number of planks to determine the width of the planks at each station, marking off each plank on each station with his stick rule.

I am there, at Ross's side, as he does this. I am asking questions. I am helping with the measuring. I am taking notes. But I still don't understand what is going on. And once he begins to take the patterns of the actual planks, I will become so lost as not to see a thing. My confusion is fascinating to Ross—it happens all the time. The planking appears to be a mystery, a physical conundrum. How do you draw the outlines of all the planks on the skeleton hull, those planks that narrow and widen and bend, then figure out what a bent piece will look like when it's flat?

I am convinced the process cannot be adequately described in words. Howard Chapelle, an American maritime historian and wooden boat authority who wrote *A Complete Handbook of Wooden Boat Construction,* has described the process of planking a boat, and Ross has actually tried to read it. "Chapelle does not know how to plank a boat" is Ross's evaluation of Chapelle's planking description.

Ross says the most reader-friendly, clear description of boat building is Bud MacIntosh's *How to Build a Wooden Boat.* I tried no fewer than twenty times to head-butt my way through his descriptions. Not gonna find out there. Not available. It can't be described in words. You have to *do* it. You have to see it. You have to lift a pattern off a boat, put it down flat on a board, cut that shape out, and see for yourself how it bends around the hull of a boat. And then do it again and again.

"After you plank a few boats," Ross explains, "you can look at the lines that you've lined off, and you know which planks are going to be S-shaped and which planks are going to curve the other direction. And until you've sprung enough planks on a boat—well, I'm only just getting really good at it, after twenty years of doing this."

But, in almost the same breath, Ross will go on about how easy it is, about how any child can do it if someone's there to show him how. "It looks like a mystery but every process is simple, just plain simple," Ross says. "There's nothing difficult about it once somebody shows you how it's done."

Both Nat and Ross use almost the same words to assess the situation: "It's very difficult to write about."

And both will admit that it can look to the uninitiated like a complex, even mysterious process. They joke about "the smoke and mirrors" of the work, about "the *mysteries* of boatbuilding," because it's so simple to them. In a way it's like an optical illusion that, once seen, can't be ignored.

When Ross has the important lines marked on the stations, Bruce and Bob begin to fasten the ribbands. One line gives him trouble, and he finds himself moving a ribband around to get it just right—the top of the turn of the bilge, one that's fair and will also complement the line at the bottom of the turn.

Ross drops to his hands and knees and sweeps away nails and sawdust and ribbands, saying, "I'm going to check this on the floor." Then he stands back up and looks at the line he's drawn and the sheerline and says, "I don't know why these are so far off." He and Bruce have clamped a ribband at the bottom of the turn of the bilge. They eyeball it; then they squat and peer along its length. "That has to go up, and this one has to go down," Ross says.

Ross crawls around on the floor some more, beneath the boat, nails in his mouth, and then scampers out in front and first tips upside down to look and then spreads his feet apart and looks at the boat through his legs.

Ross continues to eyeball the lines, talking aloud to himself— "We've got the top edge and the low edge of the sheerstrake . . ."— while Bruce drills holes in a ribband, clamps it in place, then dips screws in marine grease and drives them into the holes using a battery-powered drill.

Ross continues to mark off all the planks on the starboard side, dividing the length of each section of each station by the number of planks in that section. He uses a plastic calculator to do all the dividing. Ross says, "I love this thing." In his knotty, creased hands, the calculator looks newfangled. It reads 5.433. "I just wish it did it in eighths and sixteenths," he adds.

Ross takes the measurements from the starboard frames and simply transfers them to the port frames. He then screws a line of ribbands on that side, splicing them with a buttblock, along the low sheer. He notes that a guardrail will go on this plank, and it'll be a different color from the hull. "It will be the most distinctive line you'll see in the boat," he promises.

Ross steps back, looks at *Elisa*—she's now got some shape because of the ribbands—and says, "Look how fat this boat is when you put the ribbands on it." He grunts and says, *"Faht boat."*

He and Bruce put a third ribband on, but as it bends around the foremost frame toward its goal, the rabbet of the stem, a piece of wood is in the way. Two 2-by-10 planks have been nailed together and rest

Ross Gannon.   Nat Benjamin.

The shop and loft as seen from Beach Road, looking toward the harbor.

All photographs by Donna T. Ruhlman unless otherwise indicated.

Clockwise from top left: Robert Bennett, former buckaroo; Brad Ives, resting on a log of angelique in Suriname *(photograph by Michael Ruhlman);* Jim Bresson, caulking *Elisa Lee;* apprentice Myles Thurlow *(foreground)* and Bob Osleeb.

Assistant Caroline Chabouis *(foreground)* and Gretchen Snyder in the sail loft, above the inner shop.

Nat in the inner shop.

Rolling *Elisa Lee.*
Duane Case works the
chainfall from above as
boatwrights and spectators
steady the rising hull.

Nat, Ross, and Bernie Holzer
guide *Elisa*'s port sheer toward
them.

Nat and the 60-foot *Rebecca* in the Mugwump shed.

Rebecca as seen from below, with David Shay varnishing a new mast for the schooner *Mistral*.

Nat, Ross, and Todd McGee inspect the hull of the 45-foot sloop *Josephine,* originally named *Jane Dore IV.*

Hauling out *Josephine,* with Nat working the winch and employee David Grey at the bow.

Vineyard Haven Harbor's great schooners *(from left to right): When and If, Alabama, Shenandoah,* with the 37-foot *Venture* in the foreground.

on edge on blocks, here as on the port side; these support the weight
of the boat. Ross needs the end removed because it's sticking out too
far. It's 4 inches thick, really too deep to take off with a Skilsaw,
though a chain saw would go through it in a snap. Ross looks around,
spots a big old handsaw on the lofting floor—it's sometimes referred to
here as a cordless—and in moments his arm is pumping like a steam lo-
comotive and doesn't stop until the unwanted wood drops to the floor.
Ross bends the ribband around and drives a screw through it, fixing it
to the stem.

✸

At one-thirty Ross hops into the "Pepe's" pickup. When it doesn't
start, he hops out, finds a metal pole, opens the hood, and, sliding the
pole deep into the truck's innards, gives the starter motor several firm
thunks, then tosses the pole into the back of the truck, hops back in,
and revs the diesel engine. He hasn't had time to fix the starter, which
has bad brushes, but he's content with the current method of ignition,
in conjunction with the light switch dangling to the right of the steering wheel connected to the solenoid, and the toggle switch that turns
on the glow plugs; still another electrical cord, with a male end, hangs
out of the front grille, attached to a block heater that lets him start the
engine even after the most frigid nights. He might have gotten rid of
this old truck by now if that diesel engine weren't so efficient.

The truck is like his toolbox; in fact, they're both versions of the
same thing: Ross himself. On the seat beside him are an old shirt and a
socket set. A bolt of steel cable is tipped over on the floor on the pas-
senger side, along with a coffee can attached to a pie tin (a bird feeder
made years ago by Ross and Lyle for Jane, Ross's mom, who lives now
in Vineyard Haven; she wants it fixed). Duct tape holds the glove com-
partment closed. On the dash are two face masks (useful when sanding
lead paint off a hull), a ripsaw, a broken wooden toy gun, and a roll-up
tape measure.

Ross backs out of the gravel lot and heads up State Road toward
the old Ag Hall fairgrounds, stopping to pick up a cup of mushroom-
barley soup and a multigrain roll from the Black Dog Restaurant. He
rolls past Alley's General Store, where one can buy a flannel shirt, a
quart of milk, and a mop and rent a video—a real general store, with

unfinished hardwood floors—past the West Tisbury Congregational Church, a classical, white-sided structure with a tall steeple, a green lawn, and a white picket fence at the corner of Music Street, and turns right, into the old fairgrounds. About a hundred grassy yards down, Brad Ives, Ted Okie, and Myles Thurlow are deconstructing part of a 150-foot-long open structure that used to shelter the stalls for the live-stock at the county fair every August. "Oh, look at that," Ross says when he's close enough to see what's been removed so far. "Some nice plywood. Looks like five-eighths. I like that a lot." There is now a new Agricultural Hall and fairgrounds, so this structure is no longer used. G&B will scavenge some ply, good 28-foot trusses, and electrical wire.

A light, gusty rain has begun, and Brad, Ted, and Myles look cold and wet. Ross finishes his soup in the truck and climbs onto the roof, attacking the shingles and nails, which fall in a shower into the yard's pickup; everyone else moves faster to keep up.

At 2:25 Ross descends through the trusses, drops to the ground, and is soon barreling toward the West Tisbury School on Old County Road to pick up his son. Every Thursday he brings Lyle back to the boatyard for a couple of hours. Lyle, a healthy butterball of a nine-year-old, with shaggy blond hair, freckles, large, white teeth, and a round face, builds things—sailboats, devices for firing rubber bands—while Ross finishes work. The two then return to the small, unfinished house in the woods where Ross now lives.

He began building it on a lot he owned a couple hundred feet down from his first house, as a guesthouse, but when he and Suzy split, Suzy stayed in the house Ross had built for the two of them, and Ross had to hustle to get the new one livable. He recently got the plumbing working (obviating chilly trips to the outhouse he'd put in), but it can be cold inside in the winter because he hasn't laid the hardwood floors yet; what he has now is sturdy but temporary flooring, and you can see the ground here and there through cracks. The two-story house is 24 feet square. The dining room and kitchen are one and contain an old couch and a dining table Ross built. The kitchen counter is made of beautiful, heavily varnished mahogany, fine as any yacht brightwork. A freestanding column of bricks, a raised hearth and chimney, faces the den, where Ross has recently finished building shelves. Upstairs are a bedroom and what will eventually be another enclosed room, but isn't

yet. There's still much to do elsewhere, too. The bathroom, with its
claw-footed tub, isn't fully enclosed—you can see into it from the den
downstairs. The exterior is covered with tarpaper, waiting to be shin-
gled. Ross has plans for a huge wraparound porch, but now there's
nothing but mud off the makeshift front steps, and the general white-
trash appearance of the place is beginning to bug Kirsten, who's been
spending increasingly more of her life here.

Despite its work-in-progress state, the house is nothing but warmth
and honesty, and it's only after a while that I realize why this is. The
house feels old, as if it had been part of these woods for half a century
already. Look at those old mullions, the wavy, imperfect panes they
hold; they don't make windows like that anymore. The bricks of the
fireplace are old, and so is the granite mantel, which Ross and Kirsten
found when walking near a quarry in Maine. The main posts some
years ago supported *When and If* during her repair. This is a new house,
but Ross has built it out of used pieces he's been able to buy or scavenge from teardowns across the island over the years. The house really
is old if you measure it by the age of its parts, and so its soulful antique
warmth feels genuine. It's new construction and restoration at the same
time.

In the same way that Ross's toolbox and his truck are unvarnished
images of his interior makeup, his house is likewise simply a kind of
hologram of his being. From the beginning his actions seem to have
been driven by a fundamental compulsion to take things apart and put
them back together better than they were. There's no way to set about
making this your life's work, no course of study, certification, or se-
quence of events that leads to it. You simply begin doing it because
you can, because it's what your body loves, and eventually it becomes
your work and livelihood through a kind of willed inevitability.

✳

Ross was born in 1947 in Michigan, like Nat the youngest of
three. His father's work, as an executive for the company that made
Diamond Crystal salt, required the family to move several times, but
Ross spent his formative years in Darien, Connecticut. It was, he says,
"the best childhood anybody could hope for."

Ross was the classic all-American suburban boy, enjoying high

school for what it was and working summers at the yacht club in
Darien, where, under the guidance of a high school math teacher, he
helped to maintain and repair boats, still mostly wooden ones. His father taught him how to sail in a fiberglass boat but eventually bought a
wooden powerboat for the family to cruise in. When Ross graduated
from high school in 1964, he had no real desire to go to college, but
having no convincing alternative to propose, he gave in to his parents'
wish and attended North Carolina State University in Raleigh.

If he had to study something, it was obviously going to be mechanical engineering, given his inclination to take things apart to see
how they worked. He'd been that way practically from birth, and his
brother, too. It wasn't just a matter of Ross's dismantling a broken
clock as a boy—he took *everything* apart, figured out how and why it
worked, and then put it back together. It went beyond clever boyhood
curiosity, becoming a real aggravation in the house because too often
the contraption being dismantled was exactly what Mr. or Mrs. Gannon needed. Ultimately they forbade Ross's activities.

Ross got on fine in college, despite the fact that in the mid-1960s,
in the conservative South, most people didn't wear jeans to class, the
way Ross did. Nor were they amused by that hippie ponytail. And so
for a while Ross was an outcast. It didn't bother him, though, and in
time he came to be regarded as a friendly adversary among his fellow
students.

Ross was classified 1A but had no intention of fighting in Vietnam. He lacked the funds for grad school so instead enlisted in a program that trained German shepherds to sniff out land mines. This was
interesting work, and he was glad to be fulfilling his service requirements in a defensive capacity. He turned out to be good at it and was
chosen, along with one other young colleague, to travel to Okinawa to
train marines to work with these dogs.

Ross had been in Okinawa a month when his father, age fifty-six
and recovering from heart surgery, grew ill with an infection where a
plastic valve had been inserted into his heart. The illness proved fatal.
Ross flew home, helped to straighten out family affairs, and eventually
headed back to North Carolina to continue working in a civilian capacity. It was summertime, and he would travel through the soupy heat

to an air-conditioned office to write reports. He hated being in an office. He was adrift and miserable.

When a friend invited him to sail from Connecticut to Martha's Vineyard for a Fourth of July celebration, Ross accepted, and he had such a good time, loved the island so much, that he saw no reason to return to North Carolina, other than to quit his job and pack his bags. He'd been rebuilding a little Porsche that was half finished but drivable. He sold it to a friend who wanted it to drive to Mexico to pick up some mescaline (the friend promised to pay the balance due with the profits and was as good as his word), bought an old pickup with the proceeds, loaded up his belongings, and drove north. His Porsche days were over—he's never not owned a pickup truck since.

Ross found a place to live and began work as a free-lance carpenter, teaching himself as he went along. In no time he had an employee. And shortly thereafter, when the Steamship Authority called for bids to remove a building where it wanted a parking lot, Ross came in lowest and found himself in the business of taking things down as well as putting them up.

Most everyone who lived year-round on the island in those days knew Ross. The look of him alone was memorable—the big ponytail, the chiseled nose and fierce blue eyes, the brilliant smile flashing from within the bushy beard. To boys and men younger than he was, he seemed bigger than life. This guy moved houses. He jacked them up and pushed them hundreds of feet back from eroding shores.

Ross tore houses apart. It was dirty work, and heavy lifting, but Ross was good enough at it to make money. He formed a ragtag crew to tear things apart. His friend Mike Carroll remembers that it was almost as if Ross wanted to see who could do the most daring work with a lit chain saw in hand, like leaning off a rafter to take down an adjacent beam that was . . . just out of reach. For Ross it was a matter of getting the thing done—*If you won't do it, gimme the damn chain saw*—and he'd haul himself up and *do* it. You could say what you wanted about why, but Ross got a job done before the people who were supposed to do it had even finished discussing it.

For all the muscle he brought to the work, though, all the problem solving by force, it wasn't gross destruction. It was planned and careful:

Ross knew he could use this wood that clattered to the ground in a heap. The stuff most people sent off to the dump—old wood from old Vineyard houses, pincushions of rusted nails—was, if you just looked at it for a moment, remarkable first-growth wood from prewar construction. So Ross would tear things down and use the good stuff—free wood of a quality all but unavailable anymore—to build new houses.

He could build almost as fast as he could take down. He'd put up at least a house a year, build it right out of his head, and all those houses would seem to be from a better age of materials and construction and craftsmanship. Ross built about fifteen houses in the 1970s, using mostly material he'd taken down himself. Other builders commonly bulldozed the stuff and sent it to the dump, but there was Ross taking the nails out of perfectly good oak and tossing the timbers into the back of his truck, grinning the whole time. It was just a matter of watching how things worked, noticing how pieces of wood came together or came apart. Everything was possible.

So Ross, and the friends he would gather to help him do the work, moved houses, dismantled them, and put them back up in better shape than they'd previously known. He made a decent living working for himself, with time off in the dead of winter to make some more money sailing charters in the Caribbean. There was nothing more telling about Ross Gannon, and the way he operated, than the name of his company: And Friends. Somehow Ross eliminated himself from the public persona of his business and emphasized everyone else. He was the natural leader, all but swinging from a rope with a blazing chain saw in hand to get that piece no one else could reach, but when "Ross and Friends" shortened itself to its most revealing form, Ross could be left alone to do his work.

In his spare time he worked on boats because he loved them, loved to sail them, and found them endlessly fascinating. His first boat was a Vineyard Haven 15, the indigenous 21-foot sloop.

The Vineyard 15—"15" for its waterline length—is a classic little sailboat that was designed by Erford Burt in 1934 specifically to negotiate the heavy currents and choppy waters of Vineyard Sound. Dozens were built over the years, and Vineyarders still look at the remaining Vineyard Haven 15s with deep nostalgia. The design was so successful, in fact, that a dozen fiberglass versions were built in the 1960s and 1970s and came complete with a sales pitch: "Trade in your old woodie, and we'll make you a glass boat with all the trimmings from your old boat." Few of the fiberglass models exist today, though three plank-on-frame 15s, in good sailing condition, still ply the sound.

When Ross bought his, a boat called *So Long,* those 15s that remained could be bought for less than their original cost of $1,000. Many were cut up for firewood, but many others survived, and Ross restored *So Long* to good condition, figuring things out as he went, just as he'd once done with his mom's and dad's appliances. When he finished, he sailed her for a couple of years, then sold her for a profit and bought a bigger boat in need of repair and fixed *that,* learning what worked and what didn't, what was broken and why, sailing her for a summer or two or three, and then selling *her* at a profit. He did this over and over, living summers on the larger boats.

＊

That was how he existed for the better part of two decades, living a working-class life, to him an idyll on this rural island—largely solitary, dwelling in his work, inside himself—until 1990, when his son was born. It was then that Ross, as one friend puts it, "decided to bite the bullet"—settle down, that is, and make a go of family life with Suzy, an ideal that ultimately proved untenable. His moving out only intensified his adoration of Lyle. Ross has sometimes said he'll never be able to have another child because he can never love anything as much as he loves this child.

Now Ross has Lyle every other weekend and sees him two days a week, while custody hearings continue. On Thursdays, when he and Lyle get home from the boatyard, Ross fixes dinner and helps his son with his homework; they may play a game, or build something, work on a jigsaw puzzle, or read aloud. Ross won't have a TV in the house, though occasionally, on one of the alternating weekends when he has Lyle with him, they'll borrow one to watch a movie, such as *Old Yeller,* for a special occasion. At 7:55 on Thursday nights, he walks Lyle up the hill back to Suzy's, returns home in the dark, reads, and goes to bed. On Tuesdays he heads over to the West Tisbury School early to be a kind of teacher's assistant, thereby wangling an extra couple of court-allowed hours near Lyle. The split between Suzy and Ross, who were never married, was a lopsided and bitter one. Suzy isn't giving a fraction of an inch that she isn't required by law to give, and Kirsten's increasing presence is gasoline on the fire (Kirsten happens to be here this week, but she's leaving in the morning for two and a half months' fishing in Alaska). To all who know Ross, Suzy, and Lyle, it's a sad and unfortunate situation for all three.

Ross rockets toward school. He knows that if he makes it before two-thirty, before the corridors become a river of students streaming out, he can meet Lyle at the door of his classroom and not lose an instant of today's five and a half hours with his son. (Bob once looked over at Ross when he and Lyle were in the shop and said, "He loves that kid so much it's *scary*.") Ross slows at the flashing yellow light and turns in to the parking lot. This is the highlight of his day. He says hello to the crossing guard, and she says "Hi, Ross" in return. Ross walks with a bounce into the school. Classes are just starting to be dismissed, but the bright, carpeted corridors are calm and orderly. He sees a few of Lyle's classmates in the hall, putting on their coats, and knows class has just let out.

He enters the classroom and scans it. He looks around some more, and then the glazed expression of someone completely lost comes over him, as if he didn't know where he was. His heart sinks. Lyle isn't here. Ross walks over to the teacher, who explains that Lyle was out sick today. Ross nods. He is cordial and calm. You wouldn't know from his expression now that anything was unusual, but when he leaves the classroom, he mutters, "That's a load of *crap*." He walks back down the corridor and says, "*That's* why Lyle said he wasn't going to see me Thursday." The teachers know, the crossing guard knows, that Ross is leaving without his son because Suzy never bothered to tell him Lyle wasn't in school.

Back at the shop, Ross pushes into Ginny's office and calls Suzy. They both stayed home sick today, she tells him.

Ross returns to the outer shop and climbs up on the aft frames of *Elisa,* knowing that he won't see his son again till next Tuesday and that there's not a thing he can do about it. He lunges into work, not the fine work of fairing but instead heavy work, twisting and bending and wrestling the ribbands around the hull, crunching them against the angelique and wrenching the bar clamps tight against them, then driving screws through them, sinking them deep. When a ribband must be tapered off to fit into the bow stem's rabbet, Ross grabs a hatchet and smashes away at it, sending chunks of spruce flying through the air. His love will depart for months in Alaska in the morning, his son is being withheld from him, and a ribband takes the blows.

✳

The following week begins slowly. Ross is out, but no one appears to know why; some think he's working at Bargain Acres, others believe he's ill at home, which seems unlikely. Ross gets a bug every now and then, but it's never reason enough for him to miss work. Bruce, Bob, Jim, and Ted fasten the rest of the ribbands on the frames—twelve of them on each side—so that upside-down *Elisa* has a distinct hull shape, and then they begin to cut frames for bending out of white oak planks planed to 1¼ inches thick. Some of the frames toward the back of the boat must be kerfed—that is, cut down the middle for 2 or 3 feet from the top—so they'll conform to the 90-degree turn of the bilge without breaking; the inner half of the kerfed frame will be several inches higher than the outer side.

Talk in the shop is enlivened by the opening of the movie *Message in a Bottle,* starring Kevin Costner as a wooden boat builder, featuring Robin Wright Penn as The Woman, and directed by Luis Mandoki. Producers from Warner Brothers were on the island early last summer, overseeing the boats they were renting: two Malabars, beautiful 42-foot Alden-designed schooners. The movie has opened on the mainland and is scheduled to be shown at the theater in Vineyard Haven this week.

Gretchen was off island and saw it first thing, mainly because she loves the Malabars so much; for her the Malabar, and particularly Jim Lobdell's *Malabar,* ranks with other great works like *David* or *The Last Supper.* She's in and out of the boatyard all day long, her trilling laughter a beneficent counterpoint to the buzzing of saws and planers, and as she strolls through the shop, she stops to chat. Bob immediately asks about the movie. She says it was surprising how little the filmmakers showed the boats actually sailing, for all the hours they spent filming them.

"How did the boatbuilders come off?" Bob asks. "Did it make boatbuilders look good?"

"Yes, definitely," says Gretchen.

By the end of the day Tuesday Bob has asked everyone in the shop, "You going tonight?" Jim, of course, is going, with Ted and maybe Chris Mullen (whose wife, Tora, refuses to pay good money for such a thing). Duane and Myrtle will be there; Ginny, too. Bob can't wait, and he clearly wants this to be a group outing, stopping just short of suggesting that they all wear their Carhartt work clothes and carry pencils behind their ears.

The movie house in Vineyard Haven is a lovely little theater with the ticket window and concessions right there under the marquee, looking out onto Main Street. You get your popcorn at a split-level door inside. The theater seats 244 people, and it's an intimate experience because of how little it has changed over the decades. At a recent showing of *The Wizard of Oz,* parents and children enjoyed the Lion's long solo until the Lion froze, mouth agape, and the entire image tipped and rolled right off the screen. An audible crash followed: the projector had fallen over. "It's a real Vineyard experience," says Bob.

Movies stay at the theater for only a day or two before moving to the theater in Edgartown and then disappearing back off island, so it's easy for Bob to marshal the G&B force. Jim and Ted and Chris arrive last, having drunk enough rum to make the saccharine story palatable, Jim chuckling down the aisle to a seat in the front row.

Costner plays a boatbuilder whose true love has died and who carries the blame for it; another woman appears on the scene to restore his heart before his fateful end. His job as a boatbuilder is all but incidental, providing only some scenic backdrop and the suggestion of a romantic spirit in Costner. The story runs along predictable lines through its "tragic" conclusion: just as Costner has forgiven himself for the death of his love, he takes the boat he built out for a sail, is caught in a storm, and drowns trying to save a woman and child whose boat has smashed on some rocks. The G&B boatwrights are mainly respectful in the darkened theater, but one scene has Costner carefully sanding a sawn frame during a quiet and solemn moment. Jim, who has been using block planes on *Elisa's* sawn frames, finds Costner and his sandpaper so hilarious that he guffaws out loud. Other than that, though, people sit through the sentiment with minimal groans. When it's over they all gather in front of the theater, in the quiet chill of a deserted Main Street.

Bob speaks first: "His pants were too white, he couldn't be a boat-builder!"

"And the shop was clean," says Duane, noting that there seemed to be not a single shaving on the shop floor.

There is much chuckling, as might be expected. The people who do the work in real life, having just seen that work as depicted by people who make movies for a living, find it rich fodder for comedy—especially the end, wherein Costner dives into the "turbulent" sea wearing a heavy sweater and foul-weather gear. (It did look good on him, though.) "And leave your boat," says Jim. "Zat's a good idea."

Bob's wife, Marilyn, looks on the bright side. Referring to the smashed boat he was trying to reach, she says, "At least it was fiberglass."

"To lose a fiberglass boat and Kevin Costner in the same scene," Ted agrees, also looking on the bright side.

Bob and Marilyn, dressed in their weather suits for biking, have a long walk home. As they head off into the cold, Bob says, "We'll have to sand the frames tomorrow. Tell Nat how it's really done."

Nat, it's really no surprise, won't see the movie. It's not that he's not interested; it's that this Hollywood depiction of his profession, the work of his life, is, according to reviews and comments he's heard, not simply inaccurate in its details but worse—it's completely contrary to what he and Ross do. He liked having Warner Brothers around, liked the director ("very clean clothes, pointy shoes"), and had hoped they'd film the story here, in his shop. At the time it had seemed like a kick, and why not, especially given that Hollywood was famous for overpaying people. But now that the movie is out, and it's clear that it's a paint-by-numbers love story, he's happy to have the work of *Liberty*'s worm gear and also her bilge pump to occupy his time. Hollywood, by its own admission, by its cumulative values and its overarching ethos, is fundamentally about artifice. That's what its essence is, that's what drives it. The venerable Mr. McCullough observed the following about Nat Benjamin: "My instinct tells me that the essence of what Nat does, of what he makes, and of him, is *trustworthiness*. Trust. It has to do with the *opposite* of artifice, the opposite of sham, the opposite of surface: *content*." In other words, a beautiful, strong toolbox that doubles as a sawhorse. Worm gear. Angelique timbers. *Rebecca. Elisa. Liberty.*

✴

Ross returns late in the morning on Wednesday and climbs up onto *Elisa's* forward frames. He left his house yesterday only to get some aspirin and return to bed. First it felt like a flu bug in his stomach; then it turned into a crushing headache and body aches. Bob looks at Ross and, alluding to the Lyle situation, says to me, "Stress."

"I feel weak," Ross explains. "I feel like I've got a hangover from what I had for two days. I feel tender."

A few minutes later he's got a saw in his hand and is plowing like a moose through some angelique frames.

While he was gone, Bob, Jim, Bruce, and Nat got in a number of steam-bent frames—the first of thirty on each side. First the boys cooked them in the steamer, and Nat would say, "Get in the cage, Bruce," because as frames get secured perpendicular to the ribbands, the hull looks more and more like an actual cage. Bruce or Jim usually worked inside. When all were ready and in position with clamps in hand, a steaming-hot frame would be pulled from the steam box, the box's door slid back in, and wedges malleted down to keep it shut tight. Nat or Bob would then slide the frame through the ribbands, and Jim or Bruce would bend it into place. Jim would lie flat on the blocks and push up on a frame with his feet to get it where he wanted it to go; from the outside, Bob or Nat would pull it toward the ribbands till it was close enough to get a clamp around it, then use the strength of a bar clamp to pull it the rest of the way, flush against the ribband. Then Nat or Bob would drive drywall screws through the ribbands into the oak frame. The work required three people—one in the cage, one up at the keel with screws and a battery drill with a screw bit, one on the floor with clamps, screws, and a drill.

"I like bending frames," Bob said. "It goes quickly. You feel like you're accomplishing something. She's bending good."

Once the majority of the frames are on, the crew can begin spiling planks and fastening them to the two different types of frames, sawn and bent. Planks finally make it a boat. And Ross will spile the first plank this afternoon, the garboard, which fits into the rabbet along the keel. "We call it spiling," he says, "but all it is is lifting the pattern. The first time you see it, it looks complicated, but it's so simple. There's no counting, there's no numbers, it's something a child could figure out." Ross, up on the keel of *Elisa,* takes the measurements straight off the frames. The garboard is an important piece of the boat, the piece of the hull that attaches to the keel, and Ross, wanting the sturdiest wood at his disposal for it, will use angelique instead of wana, which will compose the rest of the hull. He transfers the measurements to a plank more than 20 feet long, pounds a nail in at each point on each station, runs a batten along the nails, whittles a new point on his pencil, and marks the first edge of the first plank. He removes the batten and nails, then grabs a Skilsaw. Stooping over, he runs the saw along the freshly drawn line. When he pushes the saw through the end of the plank, the narrow piece he wanted removed doesn't fall to the floor. "That didn't go through?" he says. "*Doggone* it." He lowers the blade another ¼ inch and makes the cut again.

Outside, Jim is taking advantage of the fact that the steam box has been cranked up to bend some deck beams. These oak beams are 7 or 8 feet long and 2½ inches thick—hefty pieces. Jim bends these big pieces, not so much for strength this time but rather because they are available—were he to saw out this curve, he'd need to find bigger pieces of wood and attain the shape by cutting off big chunks of them. By bending it there is no waste, and the added benefit is that the continuous grain makes a deck beam stronger than one sawed off in the same shape. These deck beams run athwartship and are fastened to the sheer clamp to provide critical structural strength. They will incidentally support the deck, which is curved so that water will spill off it. One day the beams will be visible to the owner as he reclines in the bunk and gazes up.

Jim has nailed blocks of wood in an arc on the platform at the top of the steps outside the shop. He has exaggerated the bend by several inches, knowing that when released, the wood will jump back. He hefts a steaming piece of oak from the box, up the steps to his frame, and clamps the center and right end. Bob is on the ground with a clamp, the platform at chest level, ready to help. Jim braces his back against the door of the shop and pushes the left end of the beam with his feet. Bob pulls with one hand, a clamp ready in the other, but the piece of wood begins to twist in the middle, so they stop. Ross, seeing from inside the shop that they're having trouble, comes out for a look.

Bob asks, "Want me to take it off and try it again?"

"Yeah," Ross says. "Why don't you let off, Jim." They unclamp the beam. "Let's flip her over."

They reclamp the center and the far end. Jim again scoots back against the door and pushes with his boots. Bob, grunting and grimacing, says, "Come on, grab, you sonofabitch." The clamp slides over the beam, and Bob can now screw the clamp handle to force it the rest of the way, flush against the frame.

"Why don't you just put two fat screws at either end and then we'll bend the next beams around that," Ross advises.

Jim stands, hands on hips, staring at the bent deck beam. "Shoo," he says. "This is a big piece of wood to steam, eh?"

"Yeah, I'll say." Ross retrieves a mallet from the shop and returns to pound on the spots where there's some twist—*whamwhamwham!* He straightens up and says, "I wonder how it will keep its shape." Bob wonders about leaving the beams out in all this weather—should they be covered? Ross says, "I'd just let them sit right out here in the hot sun and dry out, because that's what'll make them hold their shape."

It's still winter, but today is warm and sunny, and it sounds like a good idea. The three bent deck beams will remain there for many weeks and will hold their shape well when they're at last unbolted from the platform.

Elisa progresses quickly—the island is in its quietest time of the year, and there are few distractions to slow down the work. Most of those who stop by the shop are friends. The crew are strong and have locked into a routine.

✳

Part of the routine of the boatyard is lunchtime, from one to two. Throughout the winter you can usually find most of the crew eating lunch in the boatshop. If it's very cold, they'll sit inside, near the fire. If it's moderate, they'll sit in the brighter, roomier outer shop between *Elisa* and the Bella; and on warm, sunny days, which are not infrequent throughout February and March, they'll sit along the dock like birds on a wire.

Bob Osleeb almost always brings a sandwich packed in a Tupperware container. David Shay always brings a thermos of coffee, which he drinks with his sandwich and sunflower seeds. David is the quietest of the boatyard workers. He stands well over six feet, has thick brown hair just going gray, blue eyes, and a bushy mustache, and will hover for patient hours at a stretch burning off the varnish on a *Liberty* hatch, never saying more than hello—and often just raising his eyebrows—in greeting. He's forty-nine years old and knows nothing about building boats. He does know welding, though, having learned it in prison, so when the yard needs any welding work done, he's the man.

Before he went to prison, David says, he "didn't do much of anything except sell marijuana." He had a wife and children and a house in Maine and ultimately got out of the business because it was too precarious, given his growing family. But in 1994 he was lured back by a former associate for one last deal, what was to be a favor for the associate. In fact, the man was a rat; he drew David into a sting to reduce his own sentence, and David went to jail for three years. The Federal Correctional Institution in Morgantown, West Virginia, was so clean, and the facilities so modern, that inmates called it Club Fed. What hurt David most was being away from his children. His marriage ended while he was inside, and the government confiscated his house in Maine. David was released on October 4, 1997, three years almost to the day after he was incarcerated. He spent the rest of that year in a halfway house in Boston and arrived on Martha's Vineyard on New Year's Eve. He came here because his ex-wife, Betsy, and their kids had moved here. He needed to be near his children. Betsy was a friend of Gretchen's, and of the boatyard's, and she asked Nat and Ross if David might have work here. This is his first legitimate job, and he's grateful for it. Once the weather gets better, he'll begin welding the new cradle, the heavy construction of I-beams on wheels that will slide up and down the rails and into the water, hauling boats in and out. And before the year is out, he'll have learned a good deal about building boats.

David will sit on a plank at lunchtime, beside the hull of the Bella, and it will bend under his weight. He'll pour black coffee into the thermos cup and quietly eat his sandwich and nuts. Jim will pull up a crate or a ramshackle stepladder and sit across from David, eating cold cuts and cheese purchased at the A&P. Bob and Chris and Duane will sit on the ever-bending plank as well, Chris opening a tin of sardines. Sardines are a common lunch item among the boatbuilders. Because wood is abundant here, you need only roll back the lid of the can, pick up a small wedge of wood, wipe it off, and use it as a spoon for the oily fish, eaten straight out of the container. Tasty, convenient, and inexpensive. Someone else will pass around a bag of chips. And sooner or later the conversation will come around to boats and sailing. On a day when Nat is elsewhere at lunch hour, he is, briefly, a topic of conversation as well. All are aware of and respect his skills as a sailor. David Stimson, working on a small skiff on his own at Mugwump, describes Nat's acumen simply: "He knows how to make a boat *go.*" The boatwrights respect his seamanship, which at its most elemental might be defined as the ability to get your boat and your crew to safety in dangerous conditions. And they are admiring and envious of the places and distances he's covered on the sea.

Bob says, "I heard Nat went back and found someone who'd gone overboard."

All gazes turn to Bob, but no one seems to know any more, and the shop is quiet. Duane nods as he chews. He's heard the story, he says. I ask him for details—when, where, what were the conditions?

"It was in the Atlantic, pretty far out," Duane recalls, pushing his glasses up on the bridge of his nose. "It was at night. The weather, I believe, was rough, but it wasn't a storm. Apparently the guy went over when Nat was below. So he didn't even know the guy had gone over. The guy was in the water for several hours. Nat turned the boat around and found him."

Three

The Voyage of *Tappan Zee* and Beyond

I

Nat Benjamin, age twenty-one and a little banged up after his tête-à-tête with the Fiat the week before, arrived with his sea bag in late summer 1968 in the Sovereign State of Malta, which comprises three treeless, sandstone islands sixty miles south of Sicily and about twice that distance north of Libya, in the middle of the Mediterranean Sea. He was looking for his boat. *Tappan Zee* was said to be tied up at an ancient stone wharf called Ta' Xbiex, in the capital city of Valletta. Originally a Phoenician and Carthaginian colony, Malta was captured by the Romans in 218 B.C. and fought over periodically throughout its history until the nineteenth century, when the British got their hands on it and developed it into a major naval base. This function lasted through World War II, when it became the most bombed spot in the world. In the decades after the war Great Britain more or less ignored Malta until it claimed independence in 1964, by which time its economy was ragged due to the abandonment of the naval bases.

Malta's situation in the middle of the Mediterranean Sea, however, made it a magnet for an international community of sailors. These sailors were quite a bit different from the standard Newport or Long Island yachtsmen Nat had been associated with. In baking-hot Malta he found a gathering of Europeans, South Africans, Australians, and New Zealanders, most of whom had sold all they owned in their homeland and simply sailed away. He also found his boat.

"It was an eye-opener," Nat recalls. He remembers thinking, "Whoa, what the hell do you do with this contraption?"

Tappan Zee had been sailed by its current owner only once, for a
month in the Aegean Sea a few years earlier; since then she had been
tied up here and more or less abandoned to the Mediterranean sun.
She was being looked after by an agent—a quirky Maltese crook, according to Nat—named Greengrass. Mr. Greengrass was charging the
New York owner for her upkeep but doing little on his behalf. Nat
tracked down Mr. Greengrass, who explained that the boat had been
hauled, recaulked, painted, and put back in the water. He then handed
Nat an enormous bill. Nat tried to explain to Mr. Greengrass that the
bill was the owner's business.

Mr. Greengrass said, in that strange Maltese accent, "The boat
doesn't leave until the bill is paid." Greengrass was a tough, bosunlike
character employed by the all-powerful harbormaster, and he was not
to be crossed.

Nat wrote to the owner and got to work on the boat, the likes of
which he'd never before seen. He'd sailed as a teenager in little wooden
boats (his family occasionally rented a summer place on Martha's Vineyard), and his first charter had been on a wooden boat, but his real experience came from fiberglass boat deliveries and charters in the West
Indies. *Tappan Zee* was an old, husky, plank-on-frame vessel with a
schooner rig, and Nat had to set about fixing her parts. The first priority was to make a little home for himself below before he began
work on the countless chores that a neglected boat requires. The sails
were tired and the integrity of the rigging was dubious, and Nat had
only a vague idea of how to evaluate at this stage what was safe to
sail with and what was not. The engine didn't run; the stove didn't
work. Everything—the ground tackle, the lifelines, deck and cabin top,
plumbing—seemed to need some sort of fixing, cleaning, or painting.
The navigation equipment, an old sextant and a compass, appeared to
be in working condition; he couldn't say one way or the other, though,
about the vintage radio direction finder. And so Nat taught himself
how to fix these things and began slowly to bond with the boat. And,
too, with the community—"destitute international boat people," as he
describes them.

Most of these yachties were tied up in the downmarket end of
Manoel Island Quay, and their ramshackle vessels were typically spread

with broken-down toilets and engine parts. For the most part they were in their late twenties and their thirties, many with young children, though some were in their seventies, like the couple on the boat next to his, an Australian man and his veddy prawpaw British wife, who invited Nat to tea regularly and became surrogate parents to him.

Likewise tied up here were Tim and Pauline Carr, who had bought a 28-foot, seventy-year-old cutter, a Falmouth Quay Punt named *Curlew,* for $1,500. They were twenty-six and twenty-one years old and destined to become one of the best-known blue-water sailing teams ever. Setting out from Malta, Tim and Pauline would sail the oceans of the world for decades in their old cutter, a boat with no engine and no electronic navigating devices. (The Carrs now live on *Curlew* in the waters off the Antarctic island of South Georgia, a beautiful, glacier-covered spot of earth, rich in wildlife, that they have written about in their book *Antarctic Oasis: Under the Spell of South Georgia.*)

The Carrs regarded Nat curiously. To this day they recall the "unforgettable and charismatic character" with the long blond hair and the "swashbuckling" boat, but back in 1968 they remarked to each other how intriguing it was that even though Nat had a daunting workload, he seemed to spend most of his time on what seemed to them cosmetics.

Nat thrived on these eclectic lives and stories as well as on the communal, all-for-one-and-one-for-all ethic of the wharf. If you had a mechanical problem, the person with the mechanical expertise was there to help you out. An issue with rigging could be solved by someone else. And at the end of the day all would gather on one or another's boat, or occasionally meet at a bar, Britannia Bill's, for a cheap supper and some "Lacrimae Vitis," or plonk—cheap wine. For the young Nat this combination of food and drink and people and cultures in a faraway land was heady and intoxicating.

Moreover, he wasn't just hanging out. He had a purpose, like everyone else here—a job to do each day. Fixing up an old boat so that he could sail it to America required Nat to learn how things worked; if he couldn't figure a problem out himself, there was always someone nearby who could. Time was a factor. He had to leave before winter, when the weather would grow dangerously unpredictable.

Nat went through one storm right there at the wharf when a *gregali,* the local term for a nor'easter, roared through and taught him how to hold on to a yacht. The boats there were tied stern-to along the stone wharf, and some were destroyed by the storm surge. Here again the rule was, *When it's all you can do to keep your boat, your home, from being destroyed, you learn what matters.*

Nat spent nearly two months working to get *Tappan Zee* ready for blue-water sailing. By mid-November he thought he had things pretty well set (though the engine remained erratic at best).

"We made one trial sail before we left," Nat says. "A whole gang of people hopped on, and we took a sail out of Valetta to one of the islands off Malta. Everything seemed to work out. It was a nice day. Wind wasn't blowing real hard." (Tim Carr went on that shakedown sail and recalls numerous green faces, Nat's evidently not among them.) "Boat didn't leak too much," says Nat, "and I even took a sight. The old Australian guy who was with me, he was watching me as I was working it out, and he said, 'Yeah, that looks pritty good, mite!' That made me feel pretty good—*it'll be all right.*"

They returned to the stone wharf, and Nat set about preparing to leave. Many in the group were concerned. The weather in winter was at best unpredictable and usually just plain lousy. His friends, all of whom had more sailing experience in these waters, told him flat out not to go until spring. "You're crazy," they said. "This old boat, you really shouldn't take it this time of year."

But Nat had a job—he'd agreed to deliver the thing. The owner was expecting her to show up on Long Island at some point. But more significant, Nat had received a letter from Pam, who explained that she was on an idyllic island called Formentera, on the other side of the Mediterranean, just off the coast of Spain. This was too enticing for young Nat. He wasn't going to spend the entire winter in *Malta.*

There was another problem besides the weather: Mr. Greengrass and the unpaid bill. The owner had not come through. Greengrass saw Nat working hard on the boat and loading it with provisions. The agent knew he was getting ready for more than a day sail. Every day Mr. Greengrass walked out to *Tappan Zee,* visibly anxious about what he was seeing. The situation grew tense. Nat couldn't wait any longer because of the weather, but every day Mr. Greengrass, the bizarre Maltese-African, badgered him.

"Aaah! You can't leave without paying this bill!" Greengrass would say. Nat nodded and said he understood.

Nat was looking for crew to help him get past the coast of Tunisia, up through the heart of the Mediterranean to Formentera, and so he coaxed a crazy Englishman who sailed around on a homemade boat that was tied up just down the wharf.

"Why don't you give me a hand?" Nat asked, all grin and charm. "We'll get this boat to Formentera so I can see my sweetheart. You need a little adventure. You can't be sitting around *here* all winter." The Englishman was game.

Nat added, "Oh, yeah, we'll be leaving a little after midnight."

And like that they slipped out of Ta' Xbiex, away from Malta and Mr. Greengrass in the November night, and set a course westward.

The following morning the community was shocked—*shocked*— that this nice young man had stolen away into the darkness without paying Mr. Greengrass.

✳

Winds hammered Nat and the Englishman almost immediately upon their setting out. They got such a thrashing, in fact, that Nat questioned whether they should have left at all. They bashed westward through the waves, and the wind kept coming at them and coming at them. And the waves grew rougher and rougher. The boat was leaking badly from all the smashing about, and the old sails were scarcely holding together. The Englishman was not pleased. *Tappan Zee* beat three days and nights into the wind before reaching the nearest land, the southern coast of Sicily, where she staggered into the harbor of a tiny town called Agrigento. The two young sailors were greeted like heroes.

"I don't think they'd ever seen a yacht in there," says Nat. "They don't get visited often—it's not Portofino. It was a beautiful old Sicilian town, with a big square and central café where everybody gathers. We were welcomed like royalty." The mayor and the chief of police led the fanfare.

Nat and the Englishman enjoyed the attention, but Nat knew they had a lot of work to do on the boat, what with repairing the damage and trying to fix the leaks in the hull.

After a couple of days the Englishman, having had time to think things over as he helped Nat repair the boat, finally said, "It's been really fun, Nat. *But I'm going back to Malta!*" And he was gone.

Nat, too, was reconsidering the trip—seven hundred miles over water he wasn't familiar with, in a leaky old wooden boat he'd never seen the bottom of, at the worst time of year, by himself, with no working engine, no ship-to-shore radio, nothing but a compass, a sextant, and his wits. Nat worked a couple more days on the boat, lunching and hanging around the café with the crazy Sicilians, and mulled over his situation.

"Well, *hell*," he remembers thinking, "I'm not going to spend the winter *here*. Malta was OK, but I can't spend the winter in *Agrigento*."

So he got into his boat, pointed the bow west, and plotted a course for Pam in the Balearic Islands.

A couple of days out of Agrigento he found himself off the coast of Tunisia, near enough to pick up some gorgeous Arabic music on his shortwave radio as he bombed through the Mediterranean, and for a moment he had a fine sail. But then the wind picked up again, and another *gregali* tumbled in. He made his way to the Gulf of Tunis and hove to in the lee of a little island not far from the city of Carthage. He'd wait this one out here, pump the boat—she leaked quite a bit more than he'd counted on when the wind blew—and sew the sails he'd blown out.

Sailing in the Mediterranean was considerably different, he noted, from sailing in the West Indies, where you could all but set your watch by the winds. There you knew that every day it would blow between 15 and 25 knots out of the east; you could plan your arrivals and departures almost without exception. The Med in winter was proving to be completely unpredictable.

When the *gregali* settled down, Nat set out again. When he was about twenty miles off the coast, a sirocco, a powerful southern wind, thundered off the Sahara, tearing his sails and stinging him with sand. It blew hard. Nat made some headway, but he couldn't carry enough sail to stay his course and so was blown north.

The weather could be sunny and beautiful one minute, and then out of nowhere a *gregali* or a sirocco would start shrieking through the rig. The wave pattern of this sea was inscrutable. In the Atlantic the waves were nice and big and spaced out; the Med was shallow, and so the waves were short and steep and you got a nasty chop. To windward, *Tappan Zee* slammed hard into the waves, which intensified the leaking.

And yet in a strange and beautiful way, Nat enjoyed this sailing deeply. Part of the beauty was that it *was* scary. The boat leaked, the sails blew out, and the winds pushed him off course. But what was he going to do? He fixed the problems because he didn't have a choice. He couldn't suddenly say, "Just kidding," switch the boat off, and step back onto Malta—or into Garrison, for that matter. *There he was.* Middle of the sea—him, his wherewithal, and a boat. And he enjoyed this elemental thrill. He sensed he was accomplishing something, learning something useful.

Once he got off the coast of Tunisia, and land was out of sight, he had some truly exquisite sailing and logged great distance. He'd created a fine little cabin for himself, and an efficient galley—a comfortable home. While along the coast of Tunis, he'd kept his bearings from lights on land, but now he was using *Tappan Zee*'s sextant. His noon sights were coming out well, making sense, giving him his latitude, but his morning and afternoon sights were doubtful. He had the RDF, but it was highly inaccurate and out of range. And he was occasionally hallucinating, all alone on the boat, seeing the glow of a lighthouse far off in the night and sailing toward it and sailing toward it and sailing toward it, and then the sun would rise and there would be nothing there, just the infinite stretch of sea and sky.

After he'd been sixteen days at sea, the sight of Majorca, the largest of the Balearic Islands, gathered in the distance, right on course. *Tappan Zee* reached along its southern shore and sailed into the Bay of Palma, and there was the old city and chief trading port for sugar, olives, figs, apricots and almonds, leather, pottery, silks and woolens, basketwork and jewelry, a city where the palaces of the Majorcan kings and the Moorish palace, the Almudaina, still stood. Palma was a beautiful city, made more so by the fact that Nat was sailing into it after sixteen days alone at sea. Landfall. Arriving by his own means on a small boat in a foreign country of such visual and sensual and historical riches was unspeakably exhilarating.

Nat dropped anchor in the Bay of Palma and went ashore, and the first soul he met was an Englishman named Tim Light. Nat and Tim became fast friends almost immediately. Tim had been living in Spain for several years, spoke fluent Spanish,. and knew his way around the islands. Nat had left one little international community and sailed right into another. He outlined for Tim his plan to get over to Formentera, about eighty miles southwest of Majorca. He knew from the charts that there were no good harbors there, so he asked Tim to help him find a safe mooring for *Tappan Zee* and someone to look after the boat and keep her pumped out over the winter. Nat then took a ferry from Palma to Ibiza, and from Ibiza took *Joven Dolores,* an old fishing boat–ferry, to Pam on Formentera.

✴

Formentera was a rural outpost, barely forty miles square, inhabited by fewer than three thousand people, fishermen and farmers and their families. There was no electricity, no cars beyond a mile radius from the port. Plows were pulled through the soil by donkeys. Pam had traveled there with friends and liked it so much she'd simply decided to stay. Nat and she rented, for 500 pesetas a month (about eight dollars), a tiny, two-room cottage with a dirt floor, tile roof, and wood and adobe walls. A small vineyard grew out back, and a wheat field in front. They got their drinking water from a nearby well. Two minutes' walk from their cottage lay a pristine sandy white beach.

Nat felt as if he'd stepped into the sixteenth century. The women dressed in long black dresses and shawls; the men wore handmade clothing stitched by their wives. All of them worked in the fields.

"They were the nicest, happiest, most friendly people I'd ever met," Nat says. "They'd give you anything. They'd do anything for you."

When a neighbor wanted to make sweaters for Nat and Pam so they'd be warm in the winter months, she invited them out back and told them to choose their own sheep. They pointed to a brown one and a whitish one in the flock and soon wore beautiful, oily, natural-wool sweaters.

They walked the beach and read. When their cottage life grew dull, they could wander to one of two towns on the island or hop a boat to Ibiza.

"Ibiza was the hippie capital of the world," Nat says. "We settled into this community which was really crazy. If you wanted a place to watch people, you would sit on the main boulevard in one of the cafés and watch the parade of *lunatics* wandering back and forth from all parts of the world on every kind of drug you could imagine. It was hysterical to watch this scene.

"I don't think we spent more than three or four hundred dollars in four months. Food was cheap. Cigarettes cost four pesetas a pack. The wine and brandy were cheap. It was just very easy to live there. For that period of time, it was paradise. And it was interesting because of these *crazy* people from all over the world showing up."

And yet there was a sense of being lost, too, among so many people—"mostly European," Nat recalls, "some Israelis, Africans from black Africa, a few other Americans, artists, your token drug dealers, your rogues and rascals, everybody"—among so much drug taking, then, in 1969, under the guise of spiritual questing, and with no real work to do other than reading and walking and spiritual questing, some of it genuine, some of it, well, less productive. "Those are confusing times," Nat says, recalling his search for direction and his hedonistic appetites.

✷

When spring came, it was time to begin thinking about that delivery he'd set off to do the summer before, after he and Pam finished work in Garrison and flew to Europe.

He hooked up with Tim Light in Majorca and initiated a plan to get *Tappan Zee* safely Stateside. Nat knew the boat wasn't ready for an ocean crossing; he'd need to haul out somewhere, inspect the hull, and stop the leaks. Tim had been to the Moroccan city of Ceuta, on the northern coast of Africa across from Gibraltar, and he knew a boatyard there where they could get work done cheaply and well.

He and Tim gathered a few friends for a sail—Maureen, a Scottish lass they knew, and Dick, another American—and set out on the five-hundred-mile journey. Pam stayed behind, planning to fly back to the States, meet up with Nat upon his return, and spend the summer working in New England. The weather was splendid, the winds fair, the voyage pleasantly uneventful. And they found the shipyard welcoming, with its set of rails on which to haul *Tappan Zee,* and a flamboyant Spaniard who ran the hauling with the gesticulating vigor of a mad orchestra conductor. After nearly a thousand miles of sailing, this was the first time Nat had seen her hull.

She was a simple, old-fashioned, straight-keel schooner. But other than that, he didn't really understand what he was looking at. He knew she needed work, but he didn't know if she needed recaulking, refastening, replanking, or what. A little of everything, according to the camel-driving shipwrights of the yard. The boat obviously had not been recaulked, as Greengrass had claimed, only reputtied. All the rotted caulking had to be reefed out and replaced.

Nat watched the shipwrights work, and worked himself. He noticed a forward plank that looked suspect. When he pressed it with his knife, water poured out of it. As he examined the hull, though, he found that all the planks dumped water like a sponge when pressed. "Anyway, let's take out this suspect plank," he said. He didn't like the looks of it; it didn't look solid. And so they went after it, discovering as they did so that the planks were fastened with iron nails. But they couldn't get the plank out. They hacked away at it and still it wouldn't come. It was hard and sinewy. You could wring it like a washcloth, but it held fast to the frames. One of the shipwrights explained why: the wood was cypress. Its nature was to soak up water. Nothing wrong with that. Soaked up a *lot* of water. But Nat was fascinated to see the frames and the fastenings. And the Moroccan shipwrights quickly spiled a new plank into place.

And so Nat learned about cypress, and about iron fastenings, and old caulking and new cotton caulking. And he learned about a good boatyard. This one had a planer and a band saw for power tools; the rest were common shipyard hand tools—planes, broadax, adze. It was a simple fishing-village boatyard. People did their work by hand, were generous with what they knew, and let you work on your own boat. With her garboards, stem, sternpost, and butts fastened, several new planks installed, much of her hull recaulked and puttied, and her topsides painted a deep blood orange, *Tappan Zee* was afloat again two weeks after arriving in Ceuta.

✴

The four-man crew first sailed to Gibraltar. Here Nat found a ship chandlery and made two important purchases. He'd written the owner asking for money to repair the boat, and the owner had sent $500. Nat spent $200 of it on a Heath sextant, made in England, and $50 on a Breitling wristwatch chronometer. One of the first things he did was compare the sextant readings with those from the one he had on the boat. His brand-new sextant worked perfectly. The one he'd been using, he discovered, was accurate in measuring the sun at high altitudes but way off in measuring low ones, which explained why his morning and afternoon sights on the way to Majorca had never made sense. Nat probably could have had the sextant on the boat fixed, but he felt he needed his own brand-new instrument. This might seem uncharacteristic of a sailor who put so little stock in fancy or expensive equipment—Nat almost always tended toward the ragged, the handmade, the used. But not in this case.

"This was something I wanted," he says. "It's a signature of a master mariner. You've got your sextant, chronometer, nautical almanac, and sight reduction tables—that's your license to go offshore."

This was, indeed, a very good instrument, and with it Captain Nat and his crew set out for Madeira, a group of Portuguese islands off the coast of Morocco. They would spend a few days there, they thought, and reprovision the boat before making the crossing.

The sailing was beautiful, and they boiled through the Atlantic on a starboard tack, in a fresh northwest breeze. Suddenly, without warning, the starboard foremast shroud parted. Without that wire rope, they stood to lose the mast, so Nat quickly released the sheet and came up into the wind, tacking to put the strain on the port shrouds. You don't want to lose a mast a hundred miles offshore.

Over the next few hours they limped along on a port tack while jury-rigging a new shroud with some clamps. Once it was repaired, they resumed their course, again bombing along through the Atlantic toward Madeira. But some hours later—*pop!* The starboard upper shroud on the mainmast blew. Nat couldn't believe it. *What is going on?* he thought. They repeated over the next several hours the same jury-rigging aloft they had used on the foremast, first dropping the jib and the foresail. Nat left the mainsail up to steady the boat—"otherwise you're rolling like a pig," he says.

Once this shroud was fixed as well as could be expected in a rolling sea, Nat was no longer keen to charge out into the Atlantic on a starboard tack. He examined the charts, estimated the vessel's position, and saw that they were just sixty miles or so west of Casablanca. He could put the boat on a port tack, straining the good shrouds—they hadn't broken yet, anyway—and they could be in a fairly big port soon, a port with the supplies to repair the rig properly. So *Tappan Zee* slid onto a port tack and sailed east for Casablanca.

When you arrive as a stranger in another country by plane, train, or car, you're a tourist. You have to find a taxi or a bus, and then a hotel, and then a restaurant to eat at, and all the while you encounter only those natives who are there to serve you and other tourists. When you arrive under sail, you by necessity become an immediate part of the community. You must find a hardware store, a marine supply store, perhaps a blacksmith. You talk to people on the working waterfront, you gather information, and you begin to understand this foreign land with greater intimacy and in less time than you would if you'd come by any transport other than a boat.

The crew of *Tappan Zee* found a small Moroccan yacht club and a safe harbor to secure the vessel. Nat examined the old shrouds. Before the nineteenth century, masts were held up by rope made of hemp, a vegetable fiber that was very strong and had little give. Metal rope began to appear on British ships in the 1840s, changing the way a lot of things were made (bridges, for instance). The critical step in making wire rope useful for rigging ships is the creation of the loop at the end. That's what allows you to use it. The traditional method is to form the loop by weaving the strands of the end back into the cable. This splice must then be secured and protected against the corrosion of rain and salt water. First, some form of tarred hemp is "wormed" along the spiral grooves where the strands of cable meet, to prevent water from leaking into the center of the cable. A linen cloth, the same kind of material used in bookbinding, is wrapped around the worming like a bandage, in a process called parceling. The splice is then "served," with a cotton or hemp string soaked in linseed oil being wrapped tightly around the parceling; this tacky string will eventually melt into itself and harden into a primitive form of linoleum, creating a completely waterproof seal. A shroud that is properly wormed, parceled, and served will have an all-but-permanent splice.

What Nat found on *Tappan Zee*'s shrouds, where they looped around the mast, was plastic hose. It was easy to figure out why: plastic was cheap, it was easy, and it served its purpose. But over time, apparently, condensation had built up inside the tubing and rotted the steel wire.

"There are lessons to be learned at every turn," says Nat.

Finding themselves in Casablanca, the sailors decided to see the country. Who, after all, really wanted to be in America, with all those protests going on, Vietnam raging? Nat was designated 1Y because of a torn knee cartilage, so he wasn't concerned about the draft. But neither did he really want to get all caught up in protests or civil rights demonstrations, much as he might sympathize with them. Much better to sail.

✳

The international crew of *Tappan Zee,* with new upper shrouds, set sail once again, now bound for the Canary Islands. Four days later, they anchored in Las Palmas, the capital, and rerigged her for the crossing. It had been a long trip already, and it was time to get this vessel to the other side of the Atlantic. Tim and Maureen returned to Majorca, leaving the two Americans, Nat and his friend Dick Darrow, to finish provisioning the boat. Dick was about Nat's age, a skinny, blond all-American gone hippie, with a straggly beard and a thorough enjoyment of recreational drugs. The two young men spent their last few dollars on provisions, filled the water tanks, and headed west into the Atlantic Ocean, Nat plotting a course south of Bermuda that would keep them in the trade wind for as long as possible. And off they flew, yielding only occasionally to recurring thoughts of how great a distance they must travel in so small a ship.

Conditions that season—it was by now late July of 1969—were perfect, and they boomed through the waves, pressed forward by east winds in fair weather. They mostly ran during the first days and had to steer continuously as they slid down the face of the big rollers coming off of Africa. It was hard work during that first week, but the work

made them stronger and fitter. As the days wore on, they developed a
routine that felt more natural to their bodies than their routine when
they were at anchor. They took longer shifts at the helm. They ate exactly what was necessary for the work they needed to do—far less than
they were used to eating. Their skin gradually became coated with salt,
their long, thick hair grew stiff with it, and the salt felt good to them,
natural. There were no noises other than the boiling of the sea, the
wind, and the sounds of a wooden boat. They were taking on the
rhythms of the sea. They felt elemental, almost primordial. This is
the sense at sea in the middle of an ocean. You feel as though you
might exist almost out of time. It could just as well be the beginning of
time—you move through the same wind and water that were here
when life began and that in all likelihood will be here till the sun burns
out. It's just you, wood, wind, water.

Tappan Zee held up well, proving herself dependable and seaworthy after all the work on her. The sails, abstract artworks of Nat's
impromptu stitching throughout the Mediterranean, were now holding, and the rigging was strong. When they sailed close to the wind,
they could adjust the tiller with lines and leave it for short stretches.

By the time *Tappan Zee* had crossed a thousand miles of ocean—
about a third of their intended passage—Nat and Dick Darrow had developed an easy routine with the boat, felt by then entirely at home on
her on the water. At midday, on a clear day, with the wind singing at
their back, the sun warm on deck, Nat would be down in the galley
fixing himself a sandwich, and he would not expect to hear any sound
other than that of water rushing against the wooden hull. But on one
such day he did hear something: Dick's voice—"Nat!"—strangely
high-pitched and light, as if from a distance. Nat hustled above and
scanned the deck. Dick Darrow was gone. Then, directly astern—
there he was, like a bobber. Nat threw him a life ring, but the boat was
already racing down the next wave, and Dick vanished in seconds.

With 20-plus knots of wind at his back and a significant Atlantic
swell, Nat had to change a baggy old headsail and put up a smaller
jib—and do it fast—before he could head back onto the wind. Every
second mattered, and the work felt like it was taking forever—he had
to bring down the one sail, unshackle the halyard, change the sheets,
then set the new jib. His fingers felt thick and clumsy. But once he had

the best sail configuration and had headed the boat to windward, he felt better. He didn't know how far down he'd been blown from where Dick went overboard, but he felt reasonably sure Dick was directly upwind. It was imperative now, in beating back to windward, that he steer a course reciprocal to the one he'd made from the point of Dick's falling overboard, and not overshoot him one way or another. The ocean took on its appropriate dimensions of vastness, with no markers of any kind, no guideposts, jagged and endless. Nat could only guess to make short tacks of five to ten minutes. They had to be pretty close in time and had to cover the same distance. If he went on a starboard tack longer than a port tack, he might miss Dick altogether. It wasn't a rough sea, but it was big. How long could Dick tread water? Nat would ride up to the top of a wave, tip over the crest, and slide down. But throughout all this he remained relatively calm, quietly praying that he'd find his friend. Half an hour passed, four or five tacks. An hour passed.

Dick Darrow, meanwhile, was treading water, clad only in shorts. He floated on his back to rest and conserve energy, not knowing how long he'd have to wait. Then he would look up, hoping. He saw Nat before Nat saw him. *Tappan Zee*'s masts were visible first, and so he knew Nat was close. He could only wait and pray that those two tips of mast tacking one way would turn and tack back.

After approximately an hour of beating his way to windward, Nat rose to the top of a wave, and Dick did the same—there he was, Nat spotted him. And each slid down again out of view. And then up. Nat maneuvered the boat over to his friend and extended his arm.

Reaching up, Dick said, "I knew you'd find me."

✳

Nat and Dick sailed on another thousand miles, and weeks passed. They headed north-northwest when they neared Bermuda, the cluster of islands more than six hundred miles off the coast of the Carolinas.

The sea from Bermuda all the way north to the coast of Maine and Nova Scotia is notoriously dangerous in late summer and fall. Some sailors claim that this part of the world's oceans—where the Gulf Stream, up to fifty miles wide and as many as two thousand feet deep, courses like a bloodline through the western Atlantic—sees worse weather on a regular basis than any other spot on earth. Ask East Coast sailors with worldwide experience about their worst storms, and they will almost invariably tell about getting caught between Bermuda and New England. This whole voyage had taken long enough that summer was approaching autumn as Nat and Dick crossed the ocean—hurricane season. And the season of 1969 was especially fierce, with the National Hurricane Center's monitoring eighteen different tropical storms and hurricanes in the North Atlantic between August and November, or roughly one a week. Sure enough, *Tappan Zee* sailed into the tail end of one of these hurricanes, Hurricane Debbie. Debbie's winds made the *gregalis* and siroccos of the Mediterranean feel like summer breezes. Nat dropped the sails, dug out an old canvas storm trysail that hadn't been used in decades, and set it on the foremast, and that husky old wooden schooner *rode* the waves with an easy, predictable motion. It was her beaminess, her fatness and her graceful double-ended hull, that made her so stable in a storm, Nat saw. And he'd never forget that kind of comfort in those conditions, he and Dick both by then exhausted and anxious for landfall.

When the storm had passed, they still had nearly eight hundred miles of ocean to cover and had by now seen nothing *but* ocean for more than a month. A huge freighter passed near enough to spot them and altered its course to steam their way. *Tappan Zee* was something to see, this far out, and worth a detour for the freighter, if only to say how d'you do and make sure all was well, especially given that the crew couldn't reach the old little sailboat on any radio channels. *Tappan Zee* had no communicating device other than the cupping of hands to mouth. Nat and Dick waited till the freighter was close enough to shout to. They called ahoy to the crew, and the crew of the freighter called back to them, asked if all was well, and, hearing that it was, steamed off again.

Just after Nat had watched the freighter disappear over the horizon, he was below deck. He sensed his feet getting wet. Very and suddenly wet. He dropped to his knees and tore the cabin sole apart to find out where the water was coming from. He discovered the source near the foremast, one of the most stressed parts of the hull. *Tappan Zee* had sprung a plank, and water was gushing in like a fountain. Having no engine, no radio on which to call for help, and no life raft, they had no other choice but to fix the hole fast, one of them working while the other one pumped. *There they were!* No choice but to work like mad or the boat would go down and they'd die. They dropped jib and foresails and hove to under mainsail in a confused sea. Nat found some canvas, slathered it with whatever kind of water-insoluble goop and putty he could lay his hands on, and gathered tacks and a hammer. Dick Darrow tied a line around his torso and went overboard with mask and fins. Between waves, he could go under long enough to press the sticky canvas to the hull and hammer a tack into it; then he'd come up for air, let another wave pass, and do it again, wishing to God each time he went under that they'd had a staple gun. But Dick got that canvas fixed onto the hull, and it stanched the fountain enough so they could get by with a makeshift repair job on the inside and frequent pumping, and there that canvas stayed for the final six hundred miles of ocean.

✳

Forty-one days after setting sail out of Las Palmas, and more than a year after Nat had been mowed down by the Fiat on his way to meet his ship in Malta, captain, crew, and ship ghosted tired but safe into the harbor at Newport, Rhode Island, now enveloped in fog. They could scarcely see the Brenton Reef tower. They'd been in this fog for more than twenty-four hours, but Nat had managed to take a sun and moon sight just before fog shut out the light, and had been able to set an accurate course. They anchored off the Ida Lewis Yacht Club, checked through Customs, and went ashore. Nat called Pam, who'd been waiting nervously to hear from him. Then he called his parents to let them know he'd made it. His mom, unworried and unimpressed, asked him if he had a job yet. Then he and Dick found their way to an old watering hole Nat knew from his bartender days. There at the bar was his first good captain, Christopher Fay, the man who'd said to him two years earlier, "Why don't you stay?" and thereby set in motion what would become Nat Benjamin's life on boats. Fay stared at the bedraggled, exuberant ocean vagabonds and told them he was buying them dinner.

II

Nat and Pam were reunited, and for the rest of the fall Nat delivered fiberglass boats down to the West Indies. He was not impressed by their construction, and by this point he was a knowledgeable judge of such things; occasionally bad construction or cheap materials put him in danger. It was on one of these trips that he just barely got that new Bristol 40 into the harbor before she fell apart in the water in front of her irate owner.

In Grenada that fall, a friend told Nat about a man in Gibraltar who wanted to sell his boat, an old wooden boat. Nat could afford a good-sized old vessel suitable for deepwater sailing, which was not the case for any new vessel that large. He contacted the man and asked for photographs. It was time for him, he felt, to make some money sailing his *own* boat—a boat he could care about, maintain, and depend on, rather than endangering himself in these plastic tubs. Nat received the photographs and a letter specifying an asking price of $20,000. He sent back an offer of half that, not expecting to hear back, but instead the man cabled him to say, "Offer accepted!"

With money saved from deliveries, and with the help of his brother-in-law in Manhattan, he and Pam flew to Gibraltar to find his ship: *Sorcerer of Asker*, a bullet built in 1921 by Anker and Jensen in Asker, Norway, a 10-meter-class racing boat converted to ocean racing in 1959. Nat arrived at the harbor late, so it was dark. He didn't even know if the boat would be there. He began to call out "*Sorcerer!*"— wondering what the hell he'd gotten himself into with other people's money and the last of his own. Then he spotted the boat. The former

owner's wife was aboard, along with two crew. Nat approached the
boat. One of the crew, John Evans, yet another crazy Englishman and
vagabond sailor, came out on deck, took a look at Nat, and called
down to his girlfriend, "Katie! It's Jesus Christ smoking a cigarette!"

The boat was in fact ready for Nat to take possession of, and since
there were no other crew around, he asked John Evans and his girlfriend to stay on, if they chose.

Sorcerer was narrow, fast, and ready to sail. They took her first
across the straits to that excellent little shipyard in Ceuta with the
camel-driving shipwrights and hauled out. Pam, who by now was
completely pregnant, was none too pleased about sailing to Morocco,
but she proved to be a trouper. After a successful haulout in Ceuta they
set out immediately for their old home, Formentera, where Nat had
promised to find them a comfortable dwelling near the good friends
they'd made. They both thought their beautiful rural island would be
the ideal place for Pam to give birth.

Nat chartered *Sorcerer* throughout the Med, leading a thoroughly
reckless life. John knew Nat was a spiritual soul, but he especially enjoyed
their routine of waking late in the morning and heading to the bar for a
few cocktails to start the day. There was usually something good to smoke
at hand, and hallucinogens were not uncommon. If it hadn't been for
those annoying Interpol agents who were always on the lookout for
drugs, the circumstances might have been perfect, but there the police
were, and on one occasion they even contacted a distressed Joan Benjamin back in Garrison ("He got out of that one," Joan recalls. "Didn't go
to jail, thank goodness. That blew over—don't know how, but it did").

Sailing with Nat, John recalls, had very much an on-the-bus-off-
the-bus attitude, à la Ken Kesey and the Merry Pranksters. Either you
were on *Sorcerer* or you weren't. And Nat did cut a dashing figure visually, with his thick ropes of strawberry-blond hair, a beard, and a mustache that curled at either end. Put a plumed hat on him and a rapier in
his hand, and he'd be straight out of a Dumas novel.

This time was important for Nat to go through, John says—a life
of no ties, living for the day, finding out if you're blessed. And Nat *was*
blessed—with a daughter, Jessica. He arrived home from chartering
Sorcerer off the coast of France two days too late to hold Pam's hand
during the natural birth or to cradle the tiny newborn.

The following autumn, 1971, Nat and Pam crossed the Atlantic bound for the West Indies, to continue chartering. Nat hauled *Sorcerer* in Martinique at a wonderful boatyard where itinerant yachties like him worked with the shipwrights. Nat himself worked and learned. A fifty-year-old wooden boat was a great education in boat construction and repair. In Guadeloupe, he watched an open fishing boat be built by hand. The shipwrights used axes for all the frame shaping, only axes, and he realized here that it wasn't about what kind of tools you had; what mattered was knowing how to use what was available. *Tappan Zee*, and then *Sorcerer*, were his apprentice ships, and through them Nat's concerns gradually shifted their focus from pure sailing to construction. He began to recognize and respect craftsmanship, particularly where it determined safety at sea. Nat and Pam sailed *Sorcerer of Asker* north to Newport in 1972 and happened to stop in Vineyard Haven Harbor. They rowed ashore. The island appealed to them, and the people were friendly. Pam, pregnant again, was ready to stay ashore. It was time, she decided, for the family to settle down.

✦

Nat found a job at the Martha's Vineyard Shipyard, working under an old carpenter there. But the yard had moved steadily toward fiberglass—did little work on wooden boats, in fact—and Nat didn't care for fiberglass. He worked as a house carpenter for a time, fished for cod one year, and always looked for work as a free-lance boat repairman. That lovely little sailboat called the Vineyard Haven 15 had been produced in abundance here in the 1930s, and several 15s were still around and in need of repair. Here and there he'd work on a Vineyard 18, a New York 30, a bass boat, a peapod. There were signs of hope in the harbor. *Shenandoah* sat on her mooring like a queen on her throne, built by Bob Douglas several years earlier. Douglas had also bought *Alabama,* but that boat sat rotting in the harbor and would continue to do so for decades. There was a once-fine vessel named *Venture,* but she was badly in need of repair; and now there was *Sorcerer*. Four wooden boats. That was the extent of the Vineyard Haven Harbor fleet.

That fleet didn't last long; it was reduced by one quarter when Nat and Pam sailed *Sorcerer* back to the Med in 1973. Ultimately, they sold *Sorcerer* and used the money to put a down payment on a house on

Grove Street in Vineyard Haven, a genuine house, that peculiar construction of right angles and corners. It was the first length of time since 1967 that he'd been without a boat.

Nat still felt directionless. It was one thing to be without a rudder when you were twenty-seven and free, and quite another, of course, to feel that way when you had a wife (Nat and Pam, on a lark, had been married in Westminster Abbey, London, in 1970, during the *Sorcerer* days) and now two daughters and a mortgage. He knew that he'd been truly happy only when working on boats, sailing boats, and being around other characters who liked the same things. "It was the first thing I ever did that I really felt good about, comfortable about," he says now. "That I wanted to be doing."

And so he tried to figure out a way somehow to make money in boat repair, construction, and design. He built a shop at the house and extended a little shed off that to loft small boats on. The family lived hand-to-mouth. Nat began to draw boats, to design and build his own dinghies and dories.

But how could he possibly support a family of four off wooden boats? It was the mid-1970s. Nat had spent the past seven years in another world, another century—among the North African shipwrights; among the farmers of Formentera in their dark, handmade clothes, tilling fields with donkeys and plows; among the international community of boat people; and alone at sea. Now he was back in America, with Gerald Ford as president, Vietnam winding down, long lines at the gas pumps, *Charlie's Angels* on the tube. Hollywood had descended on the Vineyard in the form of a giant plastic shark to invent the summer blockbuster. And fiberglass had gone from being the rage, the hot new material you could do anything with, to being just about all there was. Who needed wood anymore? The life of a wooden boat builder was a hardscrabble one even in Maine, the nation's center of wooden boats, where wooden boats, pleasure craft and workboats, still plied the waters in numbers big enough to matter. How could Nat possibly earn a living working on wooden boats on the little summer-resort island of Martha's Vineyard?

✸

In the summer of 1974 a sturdy man the same age as Nat, twenty-seven, with a thick, wavy ponytail and an enormous beard, arrived at

Nat and Pam's house for dinner with several other friends. Nat knew of Ross Gannon, and Ross knew of Nat Benjamin, but they'd never met in any meaningful way before that night. The dinner at the Benjamins' was a sendoff celebration for the crew who were to fly to Spain the next day, pick up *Sorcerer* in Majorca—Nat and Pam had sailed her to Europe—and bring her back to the United States, where her new owner awaited her. No more crossings for Nat—he was a homeboy now.

Ross was a homeboy, too, insofar as he carried on with his work of moving homes, tearing them down, and putting them back up, even as he also continued to pursue sailing and boats. In the late 1970s, Ross would find himself with a boat named *Urchin,* a 36-foot Casey cutter, that needed a rebuild. The only yard he could work at was Erford Burt's—Burt was the designer of the Vineyard Haven 15—but *Urchin* was too big to haul out there. So he and friends hauled *Urchin* out on the beach of Vineyard Haven Harbor. Ross could move whole houses given a few rollers and a jack, so a boat was no problem at all. He built a temporary shelter around his boat and set to work. Nat Benjamin, by now a friend—the delivery of *Sorcerer,* Ross's first Atlantic crossing, had gone slowly but successfully—was a great source of information about boatbuilding, and he often wandered down to have a look at Ross's progress. Soon Nat was bringing his own steam box to the beach and helping Ross fit new steam-bent frames into old *Urchin.*

Nat himself was building, too—his first substantial commission, a 25-foot centerboard sloop.

"James Taylor was the guy who got me going," Nat says. "I'd built some little boats, some skiffs and dories, but he was the one who said, 'Yeah, let's go ahead.'"

The singer-songwriter was a longtime resident of the island and had been a friend of Nat's for several years. Taylor described the kind of boat he wanted: a daysailer that he could handle alone and take into the shoals of Menemsha Pond as well as out into the sound, one that could carry half a dozen people comfortably. Nat built a half model, and Taylor liked it. So Nat drew the lines and got to work building the boat in his shop at home.

Ross didn't particularly love building houses, and he certainly didn't enjoy the foul work of tearing them down. He did enjoy moving enormous, heavy objects, but if it were up to him, he'd prefer those objects to be boats, not houses. And so he and Nat got to talking and dreaming of what they'd *like* to do. They'd like to run a boatyard for wooden boats, a yard where anyone could haul out and work on his own boat, just like those great boatyards Nat had known in Ceuta and on Martinique.

In 1980, when the property on the beach next to where Ross was rebuilding *Urchin* became available, he and Nat decided to lease it. At the time, Nat was completing work on Taylor's boat, his first sailboat design, which he called Canvasback, after the large North American wild duck. Ross loaded her onto his trailer and drove her from Nat's house down to the beach, where he and Nat and a few dozen others helped to push *Sally May* into the water. The event signaled the launch of Gannon & Benjamin Marine Railway.

All they needed was the railway. Nat and Ross visited a dozen financial institutions looking for money. They didn't require much—ten grand would do it. At every door they were turned down. Wooden boats were a thing of the past, they were told; this was not a sound investment, those days were gone. *Trade in your old woodie, and we'll make you a glass boat with all the trimmings from your old boat.* All this they heard repeatedly. But the interviews with loan officers had one other curious thing in common. When the final NO had been issued and good-byes were in the air, the officer, dropping his professional persona, would invariably whisper, "But good luck—I really hope you can make this work."

Nat ultimately convinced his brother to loan them the money, and they put in a railway with salvaged goods, picked up an old, unwanted winch, and built a dock the following year. In 1981, Nat and the owner's brother Hugh Taylor sailed *Sally May* up to Newport, where the first wooden boat show, created at the urging of *WoodenBoat* magazine, was to be held. The charismatic Benjamin made some influential friends there, and he sold a second Canvasback based on the evidence of what he'd arrived in. Gannon & Benjamin then put an ad in *Wooden-Boat* for the design; the magazine tended to set a very solemn and self-righteous tone then, and the ad's headline, "Faster than a Speeding Duck," caught people's attention—most notably that of a gentleman in Maine, who ordered the third Canvasback (he would name her *Speeding Duck*). Gannon & Benjamin Marine Railway had begun to establish itself.

III

That first wooden boat show was more than a fortunate circumstance for Nat and Ross. It represented the future of this small industry, and the extraordinary influence of *WoodenBoat* magazine. Any initial success that Nat and Ross enjoyed building wooden boats on Martha's Vineyard in the early 1980s must acknowledge to a considerable degree Jon Wilson and his magazine. Without that magazine there would have been no show focused exclusively on wooden boats, where Nat could display his talents to a concentrated gathering of wooden boat advocates and sell his second Canvasback. And just as the show brought together builders and sailors of wooden boats, the magazine likewise created an intellectual gathering among these people. Nat sold his third boat through an advertisement in its pages. Everyone who cared about wooden boats read the magazine with gratitude in the early days. Each current editor of the magazine can identify the first issue he laid hands on and recall the jolt of amazement and thanks he felt at its existence. *WoodenBoat* would ultimately set the standard for and become the arbiter of all matters related to wooden boats. It would pass along stories of yards and boats throughout the world; discuss issues of boat construction, concerning wood and tools and joints; and offer boat plans, boat advice, and technical problem solving. It would become the town crier of the wooden boat world, transforming what had been a largely unknown, disparate, and isolated collection of backwoods carpenters into an elite class of practitioners of a venerable and worthy craft. Because the boats in its pages were truly beautiful, the magazine had the power even to turn the wooden boat builder into an artist. Nat still says that one of the biggest effects of the magazine for him was that it legitimized and helped to explain what many of his contemporaries might otherwise have considered curious, eccentric work.

All this because Jon Wilson, this odd little nobody builder, had the idea of starting a newsletter for a dying industry, modeled after nothing less than the *Journal of the American Medical Association.*

"I know, it sounds ridiculous," Jon says. "But that is really what I felt. Here were these doctors, sharing this information that was critical. Nobody questioned it: *If you learn something about a person's body that other doctors should know, you* tell *those doctors.* Because you're saving lives. Well, what I wanted to do was exactly the same thing."

Jon at the time, 1974, was a boatbuilder in Pembroke, Maine, with a wife, a child, and a cabin-cum-boatshop where they all lived. They had lived in a teepee while Jon built their dwelling. It was rough going out in the woods, trying to live off the land and build boats, even for a healthy young couple moving into their late twenties. Jon felt isolated during the building of the three boats that were the extent of his solo building effort, but even more ominous than his own struggles was his growing realization that when he *did* venture near civilization, it was increasingly difficult for him to find people who *cared* about wooden boats. He was building boats, he realized, against the backdrop of an industry that was quickly becoming extinct.

"Altogether," he says, "just completely dying. And my first concern was, as a boatbuilder, well, if the boat industry dies, and I'm one of the practitioners, where am I going to be? But the more I thought about it, the more I realized that the way I felt about it was the way a number of others felt: that this *shouldn't* die."

And here the strangest thing happened: Jon Wilson thought he could *do* something about it. This would have seemed bizarrely out of character to anyone who'd known Jon growing up. Jon Wilson had been a misfit his whole life, a runt, the guy who didn't fit in, who hated school, got into trouble, and dropped out of college before his freshman year was half over, and now he was a small-time boatbuilder, just twenty-eight years old, with a few small boats to his name and a cabin in the woods. His life story had never been one of "I can do this"; rather, it had been one of "I don't think I can do that." To this day Jon Wilson can't explain why on earth he would think he could do anything at all along the lines of preserving an entire industry.

"I think it was more ignorance than anything," he says now, "but I just didn't have any doubt. Partly it was that it didn't matter what the scale was. If I just did *something,* it would be a good thing. It didn't have to be a raging success, just anything to help, a finger in the dike, anything to stem the tide. I could do that, and furthermore I would *love* to do that. That was the other part, that it would be *really great* to do. It would be a *great* thing to *do.* I would love to put my curiosity and my experience, such as it was, to stemming this tide."

And so began the idea of a newsletter that would connect boat-builders with one another. When a boatbuilder learned about why frames were breaking so frequently in a given hull, he could tell his fellow boatwrights about it so everyone would understand the problem and its remedy, just like the doctors with their journal.

So he had a model for his newsletter, and he also had a voice to emulate—that of John Gardner, an assistant curator of small craft at the Mystic Seaport Museum, who also wrote for and helped to edit a magazine called *National Fisherman.* Gardner was a teacher, a public speaker, a historian, a boat designer and boatbuilder, a man greatly admired and beloved for his intelligence, his benevolence, and his advocacy of small wooden boats. That was what Jon Wilson loved most—the way Gardner advocated for small wooden vessels, boats that were accessible to the amateur builder. And Jon thought, The way he's an advocate, I want to be an advocate.

"I had a fascination with yachting history and with yachts, because of the exquisiteness of the work that went into them," Wilson says. "That was what I wanted preserved more than the small boat—and I happened to be *building* small boats, but what I was really interested in was the yacht side."

Jon ran into a problem as soon as he tried to get material for his newsletter. Boatbuilders are not typically the most loquacious of people—they build boats, they don't contribute to newsletters. They aren't writers, and they don't rely an awful lot on the written word. If

Jon could have traveled the country visiting boatyards, he'd have been able to gather the information he wanted, but that wasn't possible. And in the meantime, standing in the woods in Pembroke and shouting would have elicited a greater response than he was getting by means of the written word.

So he thought, Well, then, let's do a magazine.

"I didn't know that it couldn't be done," Jon says. "And I didn't know that what I didn't know could get me into deep, deep trouble.

"I didn't know how many pages of advertising I might need; I didn't know that I would *need* advertising. I didn't know how to do a financial analysis. I didn't know how to do a business plan. The only thing I knew was that the publisher of *National Fisherman* told me that he thought maybe half its readers were interested in traditional boats as opposed to fishing. So that told me that maybe there were thirty thousand readers who were interested in traditional wooden boats, so maybe we could get ten thousand readers. *Maybe.* My wildest dreams were thirty thousand readers."

To move in his mind from a newsletter to a magazine was an easy jump for Jon, because while he'd always maintained a respectful but adversarial relationship with books, he adored magazines. Read them all, learned from them, *embraced* them. If they'd used magazines in school instead of textbooks, Jon might have been a Rhodes Scholar instead of a flunk-out. It wasn't the content so much as the vehicle itself that he seemed to understand and love. Here was a boatbuilder, after all, who subscribed to—and *read*—not only the *Journal of the American Medical Association* but also other national favorites such as *Overdrive,* a periodical for independent truckers, and the *Draft Horse Journal.*

Jon and his then-wife, Susie, sold their restored Alden ketch for eleven grand, borrowed $3,500 from friends, and published 13,000 copies of the first issue of *The Wooden Boat* in the fall of 1974. It was a collection of relatively crude photographs, a few articles, and a message to "fledgling boatbuilders with an impassioned determination to build fine wooden boats," written by the editor: "There has been, until now, no contemporary journal published which addresses itself entirely to the areas of owning, building and designing wooden yachts and small boats. To the future of these young people, embarking on a career that will bring them not riches, but satisfaction, this magazine is dedicated." Jon took those words, and boxes of issue number 1, to the Newport, Rhode Island, sailboat show that September.

He was giddily proud of his accomplishment, his magazine. This is a great magazine, he thought. *Man, this thing is going to sell out right here!*

Of those 13,000 copies, he sold 600 at the show, a third of which included a nine-dollar subscription. Jon Wilson was intoxicated by the success.

The magazine proved immediately popular, and word of mouth spread the news of its arrival. "Here is a magazine devoted to what I really care about—*look,*" many early readers felt. From the beginning people responded viscerally to its subject.

Subscriptions grew steadily as Jon, Susie (then the mother of two boys), and a staff of four put out the magazine in a Maine cabin that for six months had no electricity or running water. A photograph of Jon from those days is familiar to most readers of the magazine. He is clearly in his late twenties. A dark beard fills out his narrow face. He's seated on a log by a tree in a wool cap, a checked wool jacket, workboots—every bit the backwoods Down Easter. Behind him, at-tached to the tree, is a telephone box. He cradles the receiver to his ear, pencil in hand, notebook on knee, apparently in the middle of asking an eager reporterly question: he is the Boatbuilder-Journalist covering the industry from within the Maine woods. The image is now encased in glass beside the reception desk at the offices of *WoodenBoat,* part of the Jon Wilson iconography.

Circulation built steadily, at the rate of about 8,000 subscriptions annually, and while the magazine didn't make any money in its first few years, interest was strong enough to keep the staff charged and moving forward.

Among the chief qualities of wood, in addition to its ability to float, is its inclination to burn. In 1977, fire struck the old wooden house that *WoodenBoat* by then used for its offices (an unstable neigh-bor named Carlton eventually confessed to arson, Jon says). The editors had just finished pasting up issue 16. The building was almost completely destroyed—as were the boards for the forthcoming issue and all copies of back issues—but the photographs and manuscripts for number 16 were saved, and ultimately Wilson and company were able to reconstruct it.

Fire, according to those who have endured it, is a terrible and horrifying experience, particularly when you lose your own possessions, the work of your own hands. Part of your life burns in that fire—it's that powerful and that emotional. But Jon was not devastated by the fire. Those people who asked whether he was going to continue bewildered him. Doubt didn't flicker in his mind. They might just as well have asked him if he now intended to jump off a cliff. Indeed, he emerged from the fire like a tempered blade. Before the fire, the *Wooden-Boat* venture had been serious, but it had retained in a way the feeling of being a lark. Who were they, anyway, to think they could put out an industry-changing journal? There was so little at risk that failure was not a big concern, and neither Jon nor his few employees knew what the magazine meant to its readers. Jon describes his own attitude as *"Whatever."* The fire burned all that up, showed him exactly how important this business was, and it remains the watershed event in the history of the magazine. Its biggest revenue generator had been lost in the fire: the back issues. Jon had had no idea—this was an incredible fact, back issues. People who discovered *The Wooden Boat*'s fifth issue, or its tenth or fifteenth, were buying all the ones that had preceded it. People were *saving* this product that Jon had always thought of as being inherently disposable (Ginny Jones was among them; to this day she has in her library every single issue of *WoodenBoat* magazine, 147 of them by the time *Elisa Lee*'s backbone is set up).

"When I realized that," he says, "I realized I had to be a lot smarter about how I ran the business. It had been a kind of a wooden boat club. We were just kind of having a good time, putting this magazine out. . . . What we were missing was a kind of rigorous way of testing our ideas or vetting what we were trying to do."

After the fire—arguably the best thing that ever happened to the magazine—and Jon's unthinking conviction that he would continue to publish, all manner of providential help seemed to flood in. Jon and his staff found temporary offices and began to rebuild, changing the design of the magazine, shortening its title, creating a logo. Joel White and a fellow Mainer, the writer and wooden boat advocate Maynard

Bray, materialized to offer them a dwelling in Brooklin that might serve as their new quarters. And soon Jon hired a proper publisher so he could devote himself entirely to the editorial side and leave management and business to someone who knew about such matters. Within two years the magazine turned its first profit. Circulation was growing now by about 10,000 per year. By 1981 the company was able to purchase an extraordinary estate, with a huge brick house painted white, and a brick barn and boathouse. That same year Jon realized his dream of creating, under the umbrella of the company, a school to teach the building of wooden boats, soon to be followed by a summer sailing school. In 1984 the magazine's circulation hit 100,000. It was only then that Jon Wilson believed his magazine was a success. Indeed, it had a life of its own.

✳

Nat Benjamin had begun building James Taylor's gaff sloop at about the time of the *WoodenBoat* fire; he finished it, and helped inaugurate the G&B yard with its launching, when the magazine was well on its way to financial success and industrywide prominence. Each year thereafter, G&B put out a new Benjamin-designed boat. As the fourth was being completed, John Evans, Nat's old friend and crew from the crazy *Sorcerer* days, happened to visit the Vineyard and spotted at a hardware store a newspaper clipping about Nat Benjamin, whom he hadn't heard from in a decade—Nat Benjamin, the article said, was a boatbuilder right here in Vineyard Haven. Evans, who had become successful in the New York newspaper business in the intervening years, got in touch with Nat and asked him to consider building him a boat. Nat picked up a stick and started drawing it in the sand right there on the G&B beach. A year later, Evans's 23-foot gaff sloop, *Swallows and Amazons Forever,* was launched from the G&B rails. Then came *Liberty,* the 40-foot sloop, and after that *Lana & Harley,* at 44 feet, Nat's first schooner, a design he named Joan Ellett, after his mother. (Nat may say he didn't get along with her from the day he was born, but those youthful battles seem to have made Nat's reconciliation with his mother, herself an artist—a painter and sculptor—all the more profound.)

WoodenBoat had created a collective voice in the industry, an influence that by 1984 appeared to be ubiquitous. It would showcase builders like Gannon & Benjamin and educate and inform people about wooden boats and wooden boat technology, both ancient and new; it would describe in words, drawings, and photographs these boats' significance—their power and their beauty and the *sense* of them. It gave G&B and every other subject covered in the magazine an audience that eventually would become worldwide.

Nat's success wasn't entirely the result of *WoodenBoat* magazine, of course. His success, in large measure, resulted from the force of his personality, experience, and convictions. And perhaps from the fact that he was just plain blessed. He and Pam hadn't sailed into Penobscot Bay and chosen to set up camp in Pembroke or some other rocky Maine wilderness. Nat sat down to build boats on Martha's Vineyard, an island that was a haven not simply for run-of-the-mill celebrity in a celebrity-glutted culture, but rather for a *refined* sort of celebrity— writers, journalists, politicians, artists, a brand of intellectuals who were receptive to Nat's own artistic intelligence. His boats didn't live in a vacuum; they were on glorious display cruising Vineyard Sound, on view all summer long for some of the country's wealthiest and best-known people. Walter Cronkite would come out to say a few words at a launching. Or David McCullough. Nat would chat easily with Mike Wallace of *60 Minutes* about how not enough attention was paid to is-lands struck by hurricanes, or to those solitary sailors lost at sea, including a close friend of his. Ted Kennedy, who himself owns a wooden boat, *Mya,* remains a friend of the boatyard. Movie stars and pop singers wander through to have a look. In 1993 Hillary Clinton, introduced to Nat by a mutual friend (such are his friends), would ask him to take the First Family sailing during the family's Vineyard vaca-tion. Bill and Hillary had such a good time that they would call Nat the following summer to see if he was free to do it again. For Bill and Hill, of course he was. In turn they invited Nat and Pam to the White House.

Raised in an upper-middle-class family with roots extending back in American political history to Peter Stuyvesant and New Amsterdam, Nat apparently developed enough comfort and ease in that company to discourse easily in the privileged society of presidents and senators, celebrity journalists, pop music icons, and Pulitzer Prize–winning historians. He is furthermore articulate and convincing when he speaks about the importance of wooden boats and gaff rigs, an argument he makes with the supporting evidence of his own life from the age of nineteen.

Not only did Nat have the good fortune to find and stay on the perfect spot of earth for what he wanted to do, he also had the good luck to meet another man who had found the same thing and who shared his appreciation of the elemental appeal and fundamental sense of traditional wooden vessels. Ross Gannon would become the power-house worker, the diesel engine of the operation, with his ingenious mechanical mind and rugged spirit, and he would be happy to leave Nat to the celebrities and politicians, preferring to dwell inside the work.

And so they would form a team that seemed as unlikely as it was inevitable: Ross Gannon, a man from another century, building boats from another century; and Nat Benjamin, the designer-artist and sales-man of the wooden boat, so savvy and charismatic that he didn't just sell such boats, he sold the whole ideology of them, he sold a belief system, a way of life, a philosophy based on simplicity, wholesomeness, and self-reliance. At the same time, Nat maintained the outlaw persona that he was evidently born with, the renegade spirit that had refused all schooling and led him south to Texas and life as a cowboy, then to the Caribbean, and then to Malta, where he found his ship and his destiny in *Tappan Zee*, an old wooden boat that carried him across the ocean and became in many ways a metaphor for all that he would become, and a critical source of his understanding of and belief in the truth of the wooden boat.

Four

The Workmanship of Risk

I

Ross strides into the shop on the first day of planking all energy and grin. He rotates his cocked right arm, which feels perfect after a shot of cortisone from the doctor. "It's like magic," he says, then hops onto the port side of *Elisa* to take measurements. The first plank of *Elisa Lee* is hung on the twelfth of March, and the planks proceed steadily and smoothly without stopping. March is a mixture of snowy, cold blows that toss the big schooners in the harbor, waves crashing on the beach, and brilliant, warm days, a combination that somehow helps to make this among the most productive months of the winter.

Jim drills a hole through the keel where two pieces have been scarfed together and finishes inserting the stopwater, a piece of cypress that will absorb any water that may migrate up the joined surfaces.

Ross takes his measurements inside the shop. Planks of wana 1½ feet wide and 20 to 25 feet long, all planed to ⅞ inch, are stacked in the center of the room. Ross flips the top piece, not liking the grain or the checks, and examines the one below. The wana takes a long time to dry, and as it does so, it can split right up the middle of a plank. At about 20 feet long, this second plank appears clean and stable. He puts 1-inch-thick blocks from the scrap box by the wood stove beneath the board so he can run a Skilsaw through it right there. He finds a broom and sweeps the sawdust off it. "Oookay," he says. "Let's see how fast it's going to be. I'm going to put this against the edge of the plank that's already up there"—referring to the angelique garboard fastened into the rabbet. He sets a long, slender batten on the board, running its entire length. "I saw a little bit of a curve there, so I'm going to put that there," he mutters to himself. He drives a nail to hold the batten for a slight curve. "Sight it for fair," he says, "and just put that line down. It's so close to straight I can get away with it this way." He whittles his pencil down to more graphite and drags it along the length of the batten. At each station on the board he has written the width of the plank he's drawing; having just penciled in the top of the plank, he measures from this line the indicated distances and then connects the lines by driving a nail into the board at each station at the indicated width and pressing the batten against these nails. He crouches at the foot of the plank to sight the batten, then drags his pencil hard the length of the plank. He presses so hard, and the wood is so coarse, that he needs to resharpen his pencil halfway through.

From stations nine through two, the numbers on the plank read 4½, 4⁹/₁₆, 4⅞, 5, 4⅞, 4⅛, 4⅞, 4¾—the various widths of the plank in inches.

Jim has come in here to work on a plank, too, beside Ross; he's sawing out one he's already drawn. Ross waits for him to finish with the Skilsaw, watching. When Jim finishes one side of the plank, Ross lifts the broom and sweeps sawdust off it so he can see the other lines.

"Boy," Ross says, "you got a lot to plane off."

Jim looks at his cut, which runs about an inch above the line he's drawn. "Yah, maybe too much," he says. "I'm going too fast."

"Don't go fast because I'm looking over your shoulder."

Jim explains that he never used a Skilsaw before he came here, and it's more difficult for him because he's left-handed. He finishes the remaining cuts, lifts the board, an estimate of the eventual plank, and fixes it in two vises on the workbench that runs the length of the shop's eastern wall. He finds his jack plane, presses it down on the edge of the plank, and begins to shave fractions off the board. Walking the length of the board with both hands on his plane, he removes long ribbons of wana, so bright and clean they're almost peach-colored. Ultimately, though, he concedes how much wood he has to remove and plugs in the power planer to take the plank down more quickly to size.

Ross eventually builds a jig on the power planer to make it easier to plane at an exact right angle; he loves the way his revised tool works, and says of the planks, "They're perfect."

Each plank is made fair, then square. The plank is then clamped in place to check its fit against the preceding one; a pencil is used to mark places where it needs more planing so that the inside edges will be flush against the plank above. Depending on where the plank falls on the frame, it may need backing out—at hard turns of the hull, the flat surface of the plank needs to be slightly concave to let it rest flush against the curving frames. Then a caulking bevel will be planed out. A plank may be unclamped, planed, reclamped, reexamined, unclamped, and planed again numerous times before it achieves a perfect fit and can at last be fastened with bronze screws counterset into each frame.

Ross and Jim work the port side of the hull, and Bob and Bruce work the starboard side. When Jim and Ross get their first planks fastened, Bob, working with the lackadaisical Bruce, already feels behind, senses the pressure of Ross's speed. When Ross hears Bob tell Bruce this, he says, "Don't go faster just to catch up. Let's keep the quality the same on both sides."

Bob nods, and Bruce comes at him from the other direction, baiting him as they work.

"Don't *start,*" Bob says to Bruce. "I just got here, and you've got me on four things." To himself he says, "He sets a bit down and expects me to know where it is."

"I set it on top," says Bruce.

Bob says, "Don't forget to mark the edges."

"What edges?"

"I gotta teach you again?" Bob, to himself: "He's still a child."

Chris Mullen wanders in hefting a battery for his boat's engine, intending to recharge it. He picks two bronze screws up off the floor and asks Bob, "Are these your fastenings?"

"Are they inch-and-a-half?"

"Are they? Seems kind of short."

"He wants to leave 'em short in case it ever gets refastened," Bob explains.

Eventually, if the boat is taken care of, and not damaged by the hurricanes that are a regular feature of the Virgin Islands, where the boat will live, all these screws will be taken out, and new ones will be put in—a good thing to do for a boat this size after about thirty years. Eventually the number 14 screws will disintegrate from the salt water and the effects of electrolysis. A bigger boat, fastened with number 22 screws, might go twice that long or even longer before needing refastening.

Bruce says, "This plank's got a hole in it."

Bob examines it. "Do you want to point it out to Ross?"

"Bondo."

"How deep is it? We could put a Dutchman on it."

"Or carve a real fancy bung."

"Isn't that what a Dutchman is?"

"Yeah."

Bob shakes his head and sets about work again, fitting the garboard into the rabbet, checking it. He exhales through his nose and says, "OK, so it's got to go up a little bit."

"No," Bruce says, deadpan. "It's got to go up."

"That's what I *said*."

✷

After a week of planking, Ross is still having a ball. He usually grows bored at this stage of construction, but not this time. "It's going on so smoothly," he says.

The powerboat's hull, with its flat sides and flat bottom, is different from a sailboat's. The design eliminates the need for backing out in much of the planking. And many of the planks require little or no bevel. Ross sees this as part of the logic of a working powerboat—its design is one of efficiency, not only on the water but in the construction phase as well—and this enhances his admiration for such boats, something he realizes only from doing the actual work.

When Nat finishes the new teak lazarette hatch for *Liberty*—a fine piece of mortise-and-tenon work—he jumps in to join the planking crew. Most of the bottom planks are on by now, just before the sharp bilge turn.

"Planking's a lot of fun," Nat says. "With this many people you finish before you can get tired of it. You can get tired of fairing pretty quickly."

Bob walks up behind Nat and says, "Whaddaya think of that bottom?"

"Great," Nat says. "I think it looks beautiful."

"I tell ya, I was glad she was upside down."

"Oh, yeah, nothing like putting on a plank on your back."

"Or sanding over your face."

✹

On a clear, springlike day, Myles Thurlow, the apprentice, arrives for work, for him high school work. Myles has a diminutive stature and a quiet presence in the shop. His hair is shaggy blond, and his complexion shows a youthful patina of fuzz and blemish as well as a bright, easy smile. He's back from seven weeks' sailing in the Caribbean with Vineyarder Gary Maynard and his family on Maynard's boat *Violet*. Maynard rebuilt *Alabama*, now out in the harbor, and over the course of six thousand solitary hours of his free time, he rebuilt his own ninety-year-old Scots Zulu, a distinctive vessel with its long bowsprit and topmast, gaff rig, and hand-cranked windlass. Myles describes the numerous ports they visited during his seven weeks aboard, some filled with as many as 200 or 250 boats. Myles says they encountered two other wooden boats while he was aboard, one beautiful and well maintained, the other a shambles.

Myles spends his entire day bunging—that is, gluing ½-inch-wide wana bungs, or plugs, into all the holes in the planks, on top of the countersunk screws. Thousands of screws will fasten wana to oak and angelique frames, and a bung will be made to cover each one. When planking began, Bruce said, "I feel sorry for whoever makes these bungs." As Jim took a break one day from planking to keep up with the bunging, he sighed and said, "We should invent screws with the bungs already attached." (Jim stays home the day Myles arrives for work. He tore his right finger to shreds on the electric planer; it's become infected and filled with pus. When someone asks Bob where Jim is, Bob says, "Jim, he tried to go as fast as Ross. You go just that little bit faster and that's when it happens. Ross can do it, but mortals can't.") Some crews will have bunging parties, using cold beer as a lure for free labor and amusement during the tedious chore.

Myles doesn't seem to mind the work. He performs even the most menial task with alacrity. He simply likes to be around boats. He doesn't know why. Reading Slocum and other sailors was part of it, he says. He had a formative voyage on *Violet* when he was nine, and the thrill of it has stayed with him these six years. But beyond that he can't explain his affinity for watercraft other than to say, "I like to be out on the water. So I guess if you like to be on the water, you need a boat."

And he was, apparently, born with an affinity for wooden boats. He likes "old technology, old designs," he says, and is inclined "against the flow of technology."

"Wooden boats have an almost universal aesthetic appeal," the fifteen-year-old says, adding, "I haven't seen many beautiful fiberglass boats." This was evident to him in virtually every harbor they anchored in. *Violet* was a magnet; just about everyone in every harbor had to have a look, ask questions. Myles left the boat shortly before *Violet* headed for Panama and the locks that would raise her above the Atlantic Ocean and shoot her down and out onto the Pacific, on the other side of the continent.

So Myles is back and bunging at G&B, and he always feels grateful, if not a little disbelieving, that he is even allowed to be here. "These guys are amazing," he says softly. "I don't think there's anyone like them." When he runs out of bungs, he finds another scrap of wana left over from the planking stock and lowers the drill press with the plug cutter a hundred more times, then takes this board to the band saw in the outer shop, stands it on edge, and pushes it through the blade, bungs dropping out of it as he cuts.

Ginny raps on her window, lowers the top half, and leans out to shout, "Myles, watch your fingers!" Ginny, fretful den mother, can't stand the sight of blood or, for that matter, loose digits lying on the band saw.

✳

As row after row of planks is fastened onto the boat, the boat changes her shape and her presence in the shop. Bruce walks up the portable wooden stairs leaning against the staging around *Liberty,* looks down at the half-planked hull, and says, "The boat's getting bigger every minute." And as more planks go up, and more of the boat's lines become visible, Ross is not without concerns about the quality of the fairing. He can see the imperfection best when he takes a call in the office and looks out into the shop at *Elisa Lee* through Ginny's window—the slight waviness to one of the central starboard planks. The eye has to be trained to see such a wave. The imperfection is elusive. You can stare right at it and only after several minutes of your looking will it materialize, the way a trompe l'oeil picture can turn on you. If you change your angle of sight, it can disappear altogether. If you stand too close, it's gone. To Ross, though, this one plank is glaringly out of fair.

He addresses the issue with Bob, and Bob says, "It looked like a fair line on the bench."

"It looked like a fair line on the bench," Ross returns, "but if it doesn't look like a fair line on the boat, it's no damn good."

The reason for the discrepancy is often the bevel. Putting a bevel in can change ever so slightly the width of the plank, meaning that you can cut the plank out exactly as you've marked it on the frames, and make it fair and square, but when you plane the bevel and clamp it to the frames to check its fit, it may be $\frac{1}{16}$ inch off, and that $\frac{1}{16}$ inch, at that particular spot, gets magnified as each successive plank is fitted and you correct for the error. This results in a wave or a hump in the lines of the plank.

"They think if they can rest it flush against the plank, it's fine," Ross says. "But you have to look at the big picture."

✳

Nat meanwhile works on one of the longest planks on the boat, measuring nearly 28 feet. "It's got a lot of sweep to it," he says. "I don't know if she's going to make it." Nat unclamps it with Jim's help and carries it on his shoulder to the long workbench outside the shop to plane it. He then walks it back and clamps it in place. He puts wedges between the ribband and the planks and hammers the wedges that drive the plank tight against the preceding plank. It's tighter this time, but still not perfectly flush. Nat stares at it. He draws the pencil from behind his ear and marks the plank where it needs work. "Lot of reverse bevel here," he says, marking it again. To Jim he says, "We should check the backing out."

Jim drops to his belly and slides beneath the lowest ribband to go inside the cage. "Eet's beautiful from under here," he says.

Nat takes the plank down, carries it to the workbench, and planes it at his pencil marks. He carries the board back, clamps it again, hammers the wedges in to push it up tight to the plank above it, and steps back to have another look. He looks some more. Walks its length. Then looks still more. He says, "I wonder if I should take a little more bevel off this to close her up just a little more."

And down the plank comes for more planing. When he's got it back up and clamped in place, has inserted wedges along its length and hammered those in to press it flush, Nat scans the new long plank and says, "Well. That looks OK, doesn't it?"

He drills holes using the tapered bit, three at every frame, and Jim follows with scores of bronze screws. Bungs will follow. Another plank is on.

✳

When Ross tires of planking, he leaves the outer shop in search of a tiller. It's contained within some piece of wood either here or at Bargain Acres, and he needs to find it. The tiller is for a boat named *Aquilon,* a 45-foot double-ender owned by Don and Polly Bishop of Cape Cod. *Aquilon* is a fine boat, but more, it stands as a genuine education in buying and owning an old wooden boat. Don and Polly used to own a 38-foot Sabre, a fiberglass boat, which they liked well enough but which Polly says "didn't have those nice lines that I was used to when I was growing up—I grew up with the Vineyard Haven Fifteens. That boat didn't have any soul, any character."

Don grunts at this, recalling in an instant his boatstruck affair with *Aquilon,* and says, "This one smacked you in the face with character."

Aquilon was commissioned in 1951 by a man who smuggled diamonds for a living, but he was arrested before he could take possession of the boat—or get away on her, as the case may be—and two owners and forty-five years later, Don and Polly saw her while on vacation in Trellis Bay, Tortola, the largest of the British Virgin Islands and the home of an English couple who were looking to sell her.

Don was smote. A flurry of faxes was dispatched over the following weeks. Don and Polly sold their fiberglass boat and bought *Aquilon* for $37,000. Without a proper marine survey—they were boatstruck.

"What we did was not very smart," Don says, wiser now. "Nat did the survey, and we found out it needed a lot done," he adds—in fact, more than anyone realized even at the time of the survey. Don was just retiring from real estate, so cost was a factor, but Nat said he was welcome to help work on the boat himself to defray some of the costs, and he worked all winter. Nevertheless, once *Aquilon* was hauled on the G&B railway and Nat and Ross got in there with their dental probes, spot repair transformed into a virtual rebuild.

Don seems to remain in a kind of permanent shock over the cost. He asks me not to say how much the rebuild cost, allowing only *"substantially more than we paid for the boat."* You can see him wringing his forehead as he says those words.

It was a happy day in June of 1997, then, when *Aquilon* was relaunched with champagne, a blessing from the minister of the West Tisbury Congregational Church, and, for Don, a glimpse of a side of Ross that he'd never seen before but that sticks with him to this day.

As the boat being lowered into the water jiggled and rocked in her cradle, Don saw that the plow anchor, which hung just over the edge on the starboard side of the bow, was knocking against the trailboard, marking and denting it. Don pointed this out to Ross. Ross stopped the winch and ran into the shop. It seemed to Don that he was gone only a moment, just long enough to retrieve something. In another moment Ross had climbed up to the bow and was fastening in a bronze plate to protect the wood from the knocking metal anchor. When Ross descended, Don saw that the plate took the form of a beautiful bronze porpoise, the outline of its shapely, curved back and fins and snout an elegant addition to the lines of the boat. When Don asked where it had come from, Ross told him he'd made it real quick out of some scrap. Don remarks on this small event not because of the fact that Ross fixed a problem promptly with the materials at hand, but because Ross—whose chief characteristics always seemed to be force-plus-speed—apparently with no forethought, and quickly, rendered a porpoise with such artful refinement. Don feels that the porpoise itself is a beautiful object regardless of its function.

It was the crowning touch in the noble effort of bringing a wooden boat back to life. She was beautiful. Ginny, from the cozy confines of her office, notes that all of *Aquilon*'s original lines were preserved, and says, "She's a really pretty boat with a lovely sheer." No small praise from The Madam.

✳

Two years later, in March 1999, Don Bishop begins preparing *Aquilon* for spring. She's in seaworthy condition, and he intends to sail her down to Grenada. He will leave her there until the fall, when he and Polly will fly back down and begin eight full months of cruising. Don is nearly sixty-eight years old, and it's time for an adventure.

Among the first orders of business for Don is a new tiller. The old tiller, besides being old, did not swing up out of the cockpit. This meant that you could steer the boat only if you were seated, and you always had five feet of tiller taking up the middle of the cockpit. If they were going to live aboard this boat for eight months and maybe more, Don and Polly wanted a tiller they could swing up out of the cockpit so they might, say, enjoy lunch there without having a fence between them. And sometimes you wanted to stand up and steer from that position, for a better view as you approached your mooring, or simply to stretch your legs.

And so Ross examined the iron pieces of the current tiller—the rudder shaft and the tiller head, which fitted onto the shaft at its tapered end. Ross called Whit Hanschka, the Bishops' nephew, who once worked at the G&B boatyard and has since opened up a forge in Vineyard Haven, up State Road. Whit fabricates just about anything out of iron, from purely functional hardware to ornamental ironwork for fireplaces to beds and even the handrail leading up the steps of the up-island co-op bank. Ross and Whit had a chat about what was required in fabricating a new head, with two long, flat pieces between which the wooden tiller would be fastened, and the hinge that would allow the tiller to rise to a vertical position.

Whit finishes the piece in March, and when Don is on the island, he stops by G&B. Ross shows him the new tiller head, a substantial piece of iron a couple of feet long and weighing five to ten pounds.

Ross asks Don if he wants the tiller head galvanized, plated with zinc to prevent rust, increase the life of the tiller, and enhance its looks. As with many of Ross's questions—*Do you think we should replace this garboard, this timber that steps the mast, this stem?*—it isn't a question at all, it's a command. The question mark at the end of the sentence is a nicety, purely for appearance' sake, a cosmetic intonation of the voice.

Don doesn't recognize this, though; he only squints and shakes his head. "I don't think we need to have it galvanized." Why spend the money—does it really matter? Don thinks. Ross always seems to be finding new ways for me to spend.

Ross says, "It's gonna rust on you like crazy. You'll always be after it."

Again, Don says no, don't galvanize it.

To expedite the process of educating Don, Ross uses more overt force, though in quiet tones. Don, he says, all but placing his knotty hand on Don's shoulder, get this tiller head galvanized and we'll help you to take care of it. Don't galvanize it, let it rust, and we'll fix it—but it's not going to be priority work.

Jim sees the piece a week later sitting on the jointer and examines it curiously. "What is this made of, stainless steel?" he asks.

Ross grins and says, "It's been galvanized. The bright silver is zinc."

With Whit's metal piece in hand, Ross can now fashion a tiller, a 5-foot piece of wood that will be thick and rectangular at its base, where it attaches to the galvanized tiller head, and start out straight but then curve up about 3 or 4 inches and then down slightly, ending with a lip of a handle and changing gradually from a rectangle to an elliptical shape in cross section and moving from thick to thinner.

Ross has a pattern for it on a scroll of paper Don sent comprising several sheets held together with duct tape, a profile view of the tiller. He unrolls it on top of a scrap of ¼-inch plywood, traces the pattern onto the plywood, and cuts out the shape on the ship's saw. He then takes the pattern outside, and there, at the bottom of the steps, is a log of black locust more than 5 feet long, a foot in diameter, with just the slight sweep the tiller needs.

"I didn't have to look far," he says.

He crouches and lifts the log, grunting as he does. Ross can sound like a series of various animals when he's heavy-lifting, but the most frequent noise issues when he exhales hard through tight lips, resulting in gopher cheeks and an elephant's trumpet. He walks the massive log up the stairs and into the shop and sets it with a thud on the ship's saw. It looks like a clean piece.

"Black locust is a real contradiction," he tells me. "It's incredibly rot-resistant, but it's almost always got a rotten heart." Typically there are many inches of black rot running right up the middle of it, he continues; it grows as fast as a weed, and he's seen it swarming with carpenter ants, and yet the exterior wood is clear and so hard that screws break off in it. It is, like silverballi, classified as a legume, and in the spring, when its sweet, fragrant blossoms emerge, it begins to develop seed pods as long as snow peas. Ross wants locust for the tiller, though, mainly because it's so strong.

He starts the ship's saw and pushes the log—a small tree, really— through the blade as straight as he's able. This surface, though a little wavy, is flat enough that by lying it on this fresh cut, he can make a flat surface with successive cuts and see what's going on inside this thing. He finds knots—grainy whorls where branches have begun, weak spots—as well as some drying checks that he wants to avoid. He lays the plywood pattern on one clean surface, stares at it, shifts the pattern. He pulls his folding rule from a narrow pocket at his thigh and measures the wood. "I need two and nine sixteenths, and I gotta stay away from the heart, so it's gotta come from that," he says, pointing. He's cut this log to between 3 and 4 inches thick. He talks aloud to himself throughout the day—sometimes in complete sentences, sometimes not. "Like that," he says, holding the pattern one way, then shifting it. "Don't like that." He removes the pattern and says, "I'm going to put that through the planer and see what we've got."

He sends it through the planer four times, reducing its width a little more than a half inch in all. He stares at it. He flips it. He holds the pattern over it to see if he can avoid the knots. He draws his pencil from behind his ear and pokes it into a small hole in one knot. He holds the pattern down, fingers spread wide on it, marks a few edges of the pattern on the log, then makes those cuts. He stares. "I don't think it's going to work," Ross says. "Throw it in the woodshed and use it for a smaller boat."

Ross heads out to the "Pepe's" truck, opens the hood, raps the starter with a long pole, gets in, flips the light switch that dangles beside the steering wheel, and fires up the engine. He backs it out of the parking lot and heads up State Road toward Bargain Acres. All kinds of stumps and logs are piled up in this lot of overgrown brush, along with various boatbuilder debris—chunks of lead, an old bathtub to melt the lead in, cinder blocks, a small mountain of sawdust five years old, stacks of lumber, the trusses from the old Ag Hall livestock shelter.

Ross parks the truck and begins to poke around in the brush for logs. He finds one he likes. I've come with him and now help him lift the big log, Ross grunting and huffing like various animals and then, as we lift it onto the back of the truck, his cheeks puffing out and the elephant trumpet sounding. He takes just this one log, a little more than 6 feet of locust, and drives back to the yard. He removes 1 foot off the log with a chain saw before carrying it to the ship's saw. A thick branch sticks out of it, so he retrieves a broadax from the wall, grinds a fresh edge on it, and whacks the branch off.

He's now ready to start the whole process over again, first making flat surfaces and examining the interior of the log, then holding the pattern against it and slowly diminishing the log in just the right way so as to avoid sapwood and knots but not remove anything the tiller might need. Again there are odd, deep curves of sapwood and central knots, making it difficult to see a long, narrow, but curved tiller in this huge log. Finally, after many cuts and many passes through the planer, he has the general shape of the tiller. He stares at it for half a minute.

He says, "I don't want to give him that." He continues to look at it. "That was a futile effort." He sets it next to his first effort on the jointer, saying, "Maybe Jim needs a tiller."

Nat presently is at the metal band saw just beyond the ship's saw, welding together a bronze frame for one of *Liberty*'s skylights. "What are you working on?" he asks his partner.

Ross says, "Tiller for *Aquilon*. Nothing. A lot of firewood." Locust makes excellent firewood.

Avoiding knots in the tiller and finding long, continuous grain are not aesthetic concerns. On a boat the size of *Aquilon,* the tiller must move a big rudder through a lot of water. The iron and the wood must be able to withstand an extraordinary amount of torque. Tillers break, and when they break, they do so in rough conditions and strong currents, when you're putting the maximum force on them; in other words, when they break, they do so precisely when having a broken tiller and no steering is most hazardous. Ross doesn't take the tiller lightly. Don and Polly may never need a tiller as strong as the one Ross intends to build them through his choices and artful cutting, but if they ever do, Ross wants that strength available. You never know.

Nat says his most *discouraging* moments at sea have been when boats failed him. You can't expect any breaks from the weather or the water, but you ought to be able to depend on your boat. In an Essex boatyard Nat once saw a sign for the workers that he's thought of re-creating to hang at G&B: "Upon your decisions rest the lives and property of men."

On the way back from spending a couple of hours in Lyle's class-room, Ross stops by the sawmill and finds yet another log to heft back to the shop and up onto the band saw. Again he repeats the process of trying to find an apt tiller inside a huge log. Again he's confounded by knots. After half an hour of sawing, planing, measuring, and staring, he sets the third useless tiller on the floor and says, "*Well.* Let's go see what the next plank looks like." And he heads to the *Elisa Lee,* which is getting bigger every day.

✳

Often visitors to the boatyard will see Nat or Ross making a pat-tern out of wood for a piece eventually to be cast in bronze. In addi-tion to planking and searching for a tiller, Ross is now constructing parts for *Liberty*'s boom gallows—a stand for the boom when the boat is at rest. The boom gallows comprises two posts that rise out of the deck and have curved, forked ends to hold a 2-inch slab of teak that supports the weight of the boom. The base, where the posts will attach to *Liberty*'s deck, the posts, and the curved, forked pieces that will con-nect the posts to the teak slab are to be of bronze. The forked piece has no right angles, is all rounded. If you took a baseball bat and sawed a U shape out of the fat end so you had what looked like a giant, old-fashioned clothespin, then bent the bat just above the base at about an 80-degree angle, you'd have something similar in shape to what Ross is fabricating out of various scraps of cedar and pine. He will sand this model smooth as soapstone, and it will appear, in his hands, to be in it-self a beautiful sculpture.

This work is relaxing and meditative for Ross; he is silent, solitary, and focused as he toils, and he looks forward to the alchemy that will occur when this piece is transformed to solid bronze at the Edson Foundry in Taunton, Massachusetts.

Most bronze pieces on G&B boats are created this way; almost nothing is ordered out of a catalog. Nat doesn't design his boats so that, for instance, the sheer can contain a chock found in this or that marine supply store. The bow chock, a long slot through which the anchor line feeds, is built into the Bella's toerail and is unique to that specific sheer, and so it must be individually fashioned. Even some of the blocks, those most common of pulleys—for the Bella, one particular block must contain two wheels or sheaves—are fashioned out of wood, sanded, and sent to the foundry.

Almost nowhere in America anymore can you find a place where people put together large objects made out of natural materials, each piece of which is fashioned by hand. All of G&B's work—the patterns to be cast, the lazarette hatch, the planks, the sawn frames and massive keel timbers of *Elisa Lee,* the rebuild of *Aquilon* and, one day, perhaps, its new tiller—can usefully be called workmanship of risk, according to the definition suggested by the late David Pye, a British professor and craftsman, in his book *The Nature and Art of Workmanship.*

Workmanship of risk is, generally, the making of anything individually by hand, the creation of a product that is never exactly the same twice. Its opposite is what Pye called workmanship of certainty—broadly defined as anything made by machine, each item the same every time. Pye distinguished the two types of work using an easy example: Writing with a pen is workmanship of risk; modern printing is workmanship of certainty. A plank-on-frame hull is workmanship of risk, a fiberglass hull workmanship of certainty. Creating the pattern for a boom gallows is work of risk; transforming that pattern into bronze is work of certainty. Because the outcome of workmanship of risk is never certain, the quality of it is determined by the *care, dexterity,* and *judgment* of the worker, qualities that are unnecessary in workmanship of certainty.

Pye's book is a clever discussion of the two types of work. The distinction itself has long been obvious and may be said to have inspired an entire intellectual and design movement, the arts and crafts movement of the late nineteenth and early twentieth centuries, which began in response to the awesome—and, to many, ominous—forces of the

Industrial Revolution and mass production. Pye's language and observations, however, brought to the distinction new power and meaning
in the age of plastics and so-called high technology (the book was
written in the 1960s), and they have only gathered force as humankind
has begun to dwell increasingly in the ether world of the Internet,
where workmanship of certainty is removed even from itself, reduced
to representations on a screen.

The subject is important because, as Pye argued, "all the works of
men which have been most admired since the beginning of history
have been made by the workmanship of risk."

All workmanship of risk is only an approximation of a worker's intent, Pye said, carried out with whatever skill that worker may possess.
We can imagine a perfect result, a designer can create the perfect design, but its execution can only *approach* that perfection, like a line approaching zero into infinity. The quality of the result, according to
Pye, is judged by how near to or far from the intended design it is.
Such work is defined by two criteria, soundness and comeliness—it's
got to work, and it's got to look good, and the longer it does both, the
more durable it is, the better it is deemed to be. Good workmanship
carries out or even improves upon a design; bad workmanship fails to
do so and "thwarts the design."

Pye championed the workmanship of risk—what we tend now
to call craftsmanship, all those rare, handmade things like furniture and
Shaker boxes—but he did not at the same time condemn mass-
produced goods or mechanical regulation. For him, industrial automa-
tion was a good thing. Mass production was a good thing. Imagine, he
said, having to build every automobile by hand, fashioning each piece
as you needed it. There's not a soul at G&B who wants to make *every*
bolt and screw by hand—they all make enough of their own as it is. All
those thousands of identical 1½-inch screws that are fastening planks to
the frames of *Elisa Lee* are each one of them a gift of mass production.

Pye's book also reminds us that the machines that make the screws,
that perform work of certainty, were themselves originally created by
workmanship of risk, could not exist without it.

"In the Science Museum in London," Pye wrote, "can be seen the
first of all lead screws, which Maudslay chased for the first screw-
cutting lathe, and one of the first planers, whose bed Roberts chiseled

and filed flat. How many generations of screws and plane surfaces can those two machines have bred?" (The Maudslay Pye refers to here is Henry Maudslay, 1771–1831, the father of the machine-tool industry. He invented the metal lathe, perfected a measuring machine accurate to a millionth of an inch, and fabricated other precision tools, most of which his firm needed in its work building engines for the British Navy.)

These facts are important to reiterate because so much time has passed since the early days of mass production and the Industrial Age that we tend to take boxes of screws and buckets of nails for granted, as if they grew like almonds and we needed only to shake the tree for more.

Pye seems to have been moved to write not simply in order to lay down a stern paternal reminder to appreciate whence we came (*"When I was a lad . . . ,"* the wheezy voice admonishes), but to voice his practical, even urgent, concern that because the quantity of workmanship of risk—individually handmade things—is diminishing for obvious commercial and practical reasons, we are as a culture losing not only our capacity to perform such work of risk but also our capacity to distinguish good from bad workmanship, and thus stand to lose potentially good workmanship forever. The danger is not that such workmanship will die out—some people will always be moved to make things by hand and to own things made by hand—but rather that, for lack of standards, "its possibilities will be neglected and inferior forms of it will be taken for granted and accepted."

This certainty was once a real fear in the construction of traditional boats, one that was in the forefront of Jon Wilson's mind when he started his magazine. It was all too conceivable to him that traditional boats—not small craft, the rowboats and dories that a suburban guy clever with wood could fashion in his backyard, but rather bigger boats, requiring more advanced knowledge and heavier timbers— would cease to be made for, say, thirty or forty years, a generation. Conceivable also, then, was the scenario that when those thirty years were up, the value and sense of traditional construction would be reconsidered, people would change their minds, want wooden boats back, but there would be no one around who knew how to build good ones, and we'd have to learn it all over again, or more likely accept and take for granted poorly constructed wooden boats.

Pye's main concerns, though, seem to have been aesthetic. Workmanship of risk should continue, he argued, because it creates a range of aesthetic qualities that mechanical or regulated work, which is always ruled by the marketplace, can never achieve. And Pye further wanted to acknowledge and illuminate the worker.

"The judge, the pianist, and workman," he wrote, are interpreters. "Interpreters are always necessary because instructions are always incomplete." The workman can do with his eye what the judge does with intuition and logic, what the pianist does with intuition and ear: he or she can measure with astonishing accuracy those things that can never be specified, isolate nuances that are too subtle to be described. No law book could be complete enough to handle the specifics of every individual case; no musical score could possibly convey how long each note must hang in the air, or precisely how loudly it should sound out; no boat design could determine a single, absolute outcome of every curve. Which is why we have lawyers and judges, why some musicians are better than others and some are considered great, and why there are a pair of Malabars in Vineyard Haven Harbor, both classic 42-foot Alden schooners, of which one is powerfully built and one has exquisitely rendered lines—two very different boats from the same design.

Not coincidental, I think, is how often Pye used the "old-style shipwright" to provide examples of his ideas. "The old-style shipwright with his adze," Pye wrote, "can get a nearly true flat surface or fair curve without any apparent guide, simply by coordination of hand and eye. . . . The shipwright with his adze does not finish off the surface by removing handfuls of wood at each stroke, but in short light strokes taking off the wood in shavings."

And: "Many lives on many occasions must have depended on their timing in forging the iron work for sailing ships. A 'cold shut' weld or a weld with dirt in it could remain undetected for years and then perhaps bring down a mast, or, if in an anchor, put a ship ashore."

Pye ultimately addressed the art of such work, looking to John Ruskin, the nineteenth-century writer, a critic of art and architecture, for a definition: "Art is the expression of man's pleasure in labor," asserted Ruskin. It was precisely this pleasure—and a very particular kind of pleasure it is, making things with your hands—that Ruskin isolated as what was lacking in the new society of the Industrial Revolution. He suggested that the masses were fundamentally unhappy because first, they didn't enjoy their work, and second, they believed that they would be happier not if they found *work* they *enjoyed,* but rather if they made enough money to pursue more pleasures *outside* work, as soon as they could get away from their miserable jobs. "It is not that men are ill-fed," Ruskin said, "but that they have no pleasure in the work by which they make their bread and therefore look to wealth as the only means to pleasure."

Certainly, as Ross Gannon stands in the dark shop, alone, sanding the pattern for the boom gallows, this kind of pleasure seems very close to the surface. It is not his solitary pleasure, but rather a pleasure of a more general and independent kind, a pleasure independent of Ross, that he simply connects to. This fact becomes clear when, a week later, a black-locust tiller dangles from the overhead wood rack between table saw and planer, fastened into a galvanized tiller head, turning slightly in the air, a couple coats of varnish on it and several more to go. In the empty shop, the nearly completed piece is a compelling sight, as if blossomed from out of nowhere, grown like a seed pod from the building itself.

Nat Benjamin takes this notion of workmanship a step further, beyond the pleasure that it is: when workmanship is focused on the task of fastening planks to frames and fairing curves, it can transform and elevate the worker.

"After nearly thirty years of continuous involvement with wooden sailing craft," he writes when asked to contribute a short essay to a picture book on classic yachts, "I am more convinced than ever that a plank-on-frame vessel is the ultimate in yacht construction. Not only does this method produce an enduring vessel with integrity, heart and soul, but it also requires a process that is so ancient and noble as to inspire the builder to work above his ability, to continue challenging himself in his expression of the rarest combination of science and art."

Perhaps this is why ordinarily purposeful people find themselves, unaccountably, loitering in boatyards. Just hanging around. They can sense it here, this pleasure, this inspiration, this elevation. They can smell it. It smells like wood. Sawn cedar, damp, steaming oak, even angelique that stinks like where the cows are on the farm. They can touch it, this wood—it feels good in the hand. It's actual. It's true. They can see it bent into curves, and they can imagine sailing away on it.

II

Coffee break. The boatbuilders relax in warm spring sun, seated in a row on sagging wana planks, the inviolable ten-thirty ritual. Gretchen's maroon hatchback crunches to a halt behind the winch, and she bops out, ever cheerful, and stands in the aisle between the workbench and the sagging planks, the boatwrights seated on either side.

"Who's been over to Mugwump?" she asks, still happy, but there's an edge of concern in her voice. Nat shakes his head, as do the others. *Rebecca* has been dormant for nearly two months now, with no sign of being revived, its owner still beset by financial troubles.

Gretchen looks amazed. She says, "What's the Travelift doing there?" All here are familiar with Travelifts, of course; Travelifts are used to move large boats.

Ross says, "Oh, do they have a Travelift at Maciel Marine already?"

Gretchen says, "No, it's right in front of Mugwump."

Nat says, "A Travelift or a . . . ," and he's shaking his head as though he didn't understand.

"No," Gretchen says, "a, a, a thing for lifting . . ." And then she loses her gumption to carry on with her flagging April Fool's joke. She's mad about not pulling it off and sends an *oh well* laugh to the sky.

David Stimson, now fully employed at this yard, says, "You should have said something less believable."

Gretchen then walks the line of boatbuilders, placing a gummy fish on each boatwright's shoulder, explaining that it's a French custom. Jim nods and says, *"Poisson d'Avril,"* but he has no idea how the custom originated.

The smells of spring are in the air. Ross notes how delightful dawn's birdsong is, and the boatyard is moving inexorably toward its busiest season. Gretchen's getting busy, too, with customers' needing their seat cushions and new sails, and so she slides around on her varnished loft floor with increased urgency. She heads up today to work on the mainsail for the Bella that Duane is completing. She'll spread the 267 square feet of sail across her floor, pinning the edges down with yellow-handled picks, her glossy floor mottled with black spots from these tools. Gretchen has shaped the gaff sail so that it will fill with wind in a gentle ellipse that will mirror the curves of the boat's hull and sheer. Nat provides her with all the measurements, and she does the rest by eye.

"Math's never been my thing," she explains.

She does most stitching by machine—for all the long seams and big sail work (and other work, such as cushion covers for *Elisa's* interior) she employs her trusted Singers, notably the Singer 107 W1, affectionately known as Frankie—but the corner and reef cringles, the metal rings at the corners of the sail, she stitches by hand. Gretchen says that such things are too easily mass-produced for her to charge a lot of money for the handwork: today's sailmakers have hydraulic presses that pop the rings into the sails in less than a second. Gretchen nonetheless prefers her method, which requires not high-tech machinery but rather a needle, some wax thread, a cold beer, and a portable telephone.

Gretchen's stairway ends right at *Elisa's* bow. The powerboat has been planked up—you could crawl all over her now, you could even stroll back aft, the hull is so wide and flat there. Ross has also put some shape into the stem, first hacking off the edges with a broadax, then refining it with power and hand planes till it's got some shapely curves in it. The rest of the crew sand the hull, and sand it and sand it, and sand it some more. "Smooth as a baby's bottom," Ross requests.

The following Monday, April 5, Jim and Bruce have begun the morning sweep-up with brooms and dustpans in the outer shop, Bruce scooping up big piles of sawdust and ribbons of wana. The outer shop has changed. *Elisa Lee* is now not only planked but painted. Ross and Jim did it yesterday, Easter Sunday. Jim had struggled out of bed and come to toil on his own boat. As he sweeps up Friday's debris, he says with a grin, "We paint za boat for Easter! Ross came in here all sad because he was alone on Easter. He expected no one to be here. But I was here. 'Hey, Jim, you want to paint the waterline?' 'Oh yah, sure.' " Jim grins widely—it was the last thing he wanted to do, but he always says yes. Then he adds, "But I couldn't because I was so hung over."

He, Ted, and Chris Mullen had toured the Oak Bluffs bars the night before, and painting a straight, crisp waterline was next to impossible, so he asked if Ross would paint the waterline, and offered to take over the big paint roller instead. White primer above the waterline, red lead below.

Ginny emerges from her office at a stroll, just poking around. She says, "I got a call from Andy. He's at Fernandina Beach."

"Oh?" Jim says with interest.

"He says he was bored of motoring through the waterway. He's going to go out and catch the Gulf Stream up to Beaufort."

Andy Lyon, one of the yard's best boatwrights, has been gone since last October, cruising in his 35-foot Crocker sloop, *Harmony,* and the bosses are eager to have him back.

Ross arrives in the "Pepe's" truck, and Jim and Ted follow him into the shop, to the bench, where he tears open paper bundles of untreated cotton. "This is called a skein, just like yarn," he says. "I want you to roll each one into half a dozen balls. Find a good container."

Ross has had a hand in caulking every boat that's rolled down the G&B rails into the harbor, and it's time to caulk *Elisa Lee.* Ted and Jim will learn Ross's method of caulking.

Tools are gathered, wooden mallets and irons. A proper caulking mallet is shaped not unlike a polo mallet, with a long, narrow head. Ross says, "I didn't have an actual caulking mallet till about four years ago. You need them. On a big boat they make sense. You want all that mass directed right on the iron. But for a smaller boat I like a bigger surface area—you don't have to look as much when you're hammering." For *Elisa,* the old wooden mallets that Ted, Jim, and Bruce will use will be fine.

Ross sets a stiff canvas bundle on the bench and unfolds its flaps.

Inside are his irons. "For this little boat," he says to Ted and Jim, "you only need one size iron. The smallest one." The irons have a wide, flat edge for driving the cotton into the seam, rising into a cylindrical stem and a big, flat head at the top. The edges vary. Some are 2 inches wide, others ½ inch. Their thickness varies as well; some are thick and square, some taper into what approaches a dull blade. And some edges are convex, so that a lot of pressure can be applied to an exact point. The best irons, the old irons, are especially pleasing to look at because of their nearly black, smooth finish.

"The finish on the new ones is just terrible," Ross notes. "This one," he says, holding up an old one to prove his point, then hesitates—"well, this is not a good example because somebody left it out in the rain and it rusted and is all pockmarked." But even this ill-cared-for iron seems to carry its own sepia backdrop, appears to have been transported here straight out of the pre–World War II Machine Age.

Ross strides to the outer shop with his tools, and Jim and Ted follow carrying cotton and mallets. They climb on top of *Elisa Lee,* and Ross continues his instructions. He likes to caulk a boat. It's what keeps the water out, what makes these pieces of wood a boat. Caulking creates the ultimate dynamic in the hull. Up to this day, *Elisa*'s hull has been a loose basket; the caulking will turn it into a drum, watertight and rigid. And this takes skill. In the golden days of ship building, there were groups of itinerant caulkers who did nothing but caulk, traveling to whichever boatyards needed it done. You must develop a caulking sense, because you can put too much or too little cotton in, and you can pop the caulk through the inside, driving the planks apart. You learn how to listen to the iron; there's a particular sound when you've placed it just right and you drive the proper amount of cotton into the seam the proper distance, striking the iron perfectly—a solid sound, as when you hit a nail sweetly. The sound of a wooden mallet against iron is pleasing.

There are specific ways of holding an iron: you hold it not like a spike but like a scoop, the heel of your hand facing upward in a graceful posture. The method is more or less to scoop the cotton in overlapping tucks and chink it lightly into place, then drive it into the seam.

Ross demonstrates and talks. "Make sure, when you drive the iron, you're hitting the seam," he says. It's easy to lose concentration and put a big gash in the edge of a plank. "See how I keep the corner of the iron in the seam?" He's begun with the first three seams in the back of the boat, one away from the seam where the garboard connects with the keel. "Always set the smallest seam first, because it's going to push the others forward slightly. And caulk the garboard last."

"Is that because it's the biggest?" Ted asks.

"Yeah, it would just make that seam bigger." He thunks away, the hull echoing slightly. "With a wider seam, you put the tucks closer together."

"Ross?" Robert Bennett interrupts. He's standing below Ross, holding out his paycheck.

Without a word Ross takes it, finds a pen, and signs it against the hull, and Robert departs to begin work on *Liberty*.

Bruce has plugged a radio into an extension cord and set it atop the keel. As Ross teaches—"That seam's too small; it needs to be reefed out before you caulk it. . . . Be careful where there's a lot of caulking bevel; there's not a lot here, where it's flat, but around the bilge, remember it's always the lower edge that's square. . . . You just have to be a judge of what's right. It takes a long time to learn how to develop that motion, to be able to do it, and do it all day long. . . . I want you to develop good habits and a good way of thinking . . ."—the voice of James Taylor, the man who sparked Gannon & Benjamin into being when he commissioned *Sally May*, sings from the radio, *"Don't let me be lonely tonight,"* and slowly Jim and Ted and Bruce begin to work on various seams, first slowly, then more rhythmically. The *thunk, thunkthunk thunk, thunk thunk, thunkthunkthunk* begins to ring out through the shop and out onto Beach Road. The funky rhythm will sound out for three days before all the seams are tight with cotton and you can rap any plank and it will not vibrate and sound hollow but only thud solidly. The boat will be sealed.

✳

Robert Bennett, signed paycheck in pocket, has climbed up onto *Liberty*, where he's lately been Dyneling the deck with Chris Mullen. Occasionally you can find him out on the dock painting a little skiff or

sanding a spar, and soon he'll be over at Packer Park, sanding and varnishing the boats G&B hauled for the winter. But for the past few
months he's mostly been aboard Doug Cabral's 40-foot sloop, dismantling sections of the interior and repairing or revarnishing the wood.
He wears knee pads over Carhartt canvas pants, and a long-johns shirt;
he stands about five foot six. His hair is thick and dark blond and falls
nearly to his shoulders. He covers his round head with a bandanna or a
ball cap most days. He keeps his full beard cut short. He is forty-two
and chose a life of sailing and wooden boats and boatyards four years
ago. At the time he didn't even know how to swim.

Here is a measure of Robert. Having spent twenty years as a
cowboy, covering millions of hot, dusty acres on horseback from
West Texas as far down as the Red River and on up to Montana and
into British Columbia, lighting sagebrush fires to brand cattle, birthing
them, leading them to water, and watching after their lives, eating most
meals of his life from a chuck wagon, living a largely solitary existence
in the awesome expanses of the western United States as a buckaroo, as
such men are called, he decided to "sell the ranch" and buy a boat.
He'd spent the first half of his working life in desert and on arid plain;
he determined he would spend the second half on the water. The only
problem was that nagging inability to swim. So: he became a lifeguard.
A primary qualification for that job, of course, is knowing how to
swim, and this is apparently how Robert makes transitions or learns
things—in drastic, effective ways.

At the age of thirty-eight, Robert had seen the life of the buckaroo changing for the worse. The half-million-acre ranches he and his
fellow buckaroos covered on horseback had begun to hire helicopters
to do the work. Fences began to go up where there had been none before. The old ways of doing things disappeared. Robert trained cutting
horses, some of the best-bred horses in the world; he loved working
with these animals, training them to manage cattle, but he didn't get on
well with wealthy owners who wanted to use them only in competi-
tions, nor did he feel comfortable in that cliquey, glitzy world. He
found many of the people crass, rude, and arrogant, and so working
with horses purely for the purposes of show ultimately depressed him.
It was time to move on. After a summer and winter as a lifeguard at the

Eureka County pool in Nevada, he bought a ticket to San Juan, Puerto Rico, and, carrying a backpack containing a pair of shorts, a pair of dress pants, a couple of T-shirts, and a pair of sneakers, made his way to a port that his guidebook described as being populated by transient sailors. He met one such couple his first day there, and they showed him around the harbor, and then he met an elderly sailor who couldn't quite handle a boat on his own and invited Robert to sail with him. In this way, Robert learned how to get around on a boat.

He bought himself a boat, a wooden boat, a 30-foot Atkin cutter. He worked at boatyards and sailed around the islands. He met a Brazilian woman named Laura who was married to a man named Bruce Davies. They became friends, sailed together. Bruce found work for Robert in Haulover, on neighboring St. Thomas. And Bruce told Robert about a great wooden boat yard up in Massachusetts, on Martha's Vineyard, where he'd worked. In July 1996 Hurricane Bertha blew through the islands and wrecked Robert's boat. The island was desperate for salvage divers after the blow, so Robert one night read a book on how to scuba dive and became a salvage diver the next day.

Soon, though, he realized he was off course, and needed to return to boats in a drastic way. "If I was going to learn about wooden boats," he says, "I had to be where there were a lot of wooden boats."

And so he figured he'd give that boatyard Bruce had mentioned a try.

There was never a question about what kind of boats he wanted to work on and be around. Robert had always been the old-fashioned, traditional type; he'd chosen to be an old-fashioned, traditional cowboy, and so it would be with boats, because that was his way. It wasn't that they were the only kind he could afford, it wasn't that he could fix a wooden boat by himself if he needed to, and it wasn't that he thought that they were prettier—it was simply something he *knew*. For him he knew right and he knew wrong, and wooden boats were right.

"The thing about wooden boats for me," Robert says, began "when I started reading about boats, I started reading Slocum, and everybody since Slocum who went to sea in small boats, before fiberglass came out and all these synthetic materials. Men who built their own boats and went out and had an adventure. They didn't just have the money and go out and buy a fancy yacht; they lived it. They were men who loved life, and they lived it by building their own boats and going to sea. And I think that's the way it should be done. For me. Same as buckarooing. You depend on yourself. You don't depend on other people."

So Robert headed north, got a job at G&B, and bought a 21-foot double-ended sloop for eight hundred bucks. It's now under tarps in front of the woodshed. When Robert removes the tarps, he can look down into the hull and see the ground through all the missing planks. "She's going to be the sweetest boat in the harbor," he says dreamily.

Last October he got the opportunity to go on his first blue-water sail. His colleague Andy Lyon was planning a trip south for half a year or so. Did Robert want to come along?

Andy, thirty-four years old, had been living on his boat since he bought her, thirteen years earlier. He knew pretty much every fastener in that 35-foot vessel, which had been built around 1940. It had a laid teak deck, and every night he threw a couple of buckets of salt water on it; it would stay wet till morning. The moisture kept the teak decking fat and Andy's bunk dry, while the salt preserved the wood, much the same way brine preserves pickles. Andy referred to this task as pickling the deck.

Andy is about Nat's size, with short, dark hair and a high forehead over tiny blue eyes, a long nose and a long chin, between which he wears a dark mustache. Andy doesn't say much. He has the laconic, dry manner of a genuine Mainer, though he grew up in Connecticut, where his family sailed a wooden boat on Long Island Sound.

"I like wood because I'm a boatbuilder," he will tell you, "and fiberglass is a lousy thing to work with." End of discussion.

Andy likes to drink beer and talk and laugh, but the image that lingers in your mind is of Andy walking off alone, or Andy staring out, alone in his thoughts.

He enjoys his work and is a talented boatbuilder, which is why Ross and Nat like it when he happens to be anchored in Vineyard Haven Harbor. Andy is likely the most efficient builder in the shop, perhaps simply by virtue of the fact that he doesn't talk to anyone. Sometimes you will see him working in the outer shop wearing the ear guards that normally hang on the planer crank inside, even though no one has been using the planer or intends to. When he wears them, he can be alone in this busy shop. He gets more done that way, he says.

"I don't like being called a live-aboard," Andy says. "I don't live aboard. This is my life. I live on something I am.

"Some people live in cabins in the mountains. Some people live in cities. It's where I live, what I do for a living. I never thought much about it. Some people live on farms. I live on the ocean.

"I live on it," he continues, referring to *Harmony,* "because it takes me places. Some people live on a boat, they tie it to a mooring and they never leave it. I don't know why you would put yourself through that and not use it as a boat. It's like living in a trailer park. Why do you live in that aluminum box if you're not moving? So I live on a boat because it takes me places. I take my home with me. It's a good way to travel. It's a good way to get somewhere you haven't been. It's always an adventure. And every night, you come home. When I stop moving, that's when I'll move off. When I'm tired of being wet and salty, I'll buy a house in the mountains. And then somebody will ask me, 'Why do you live in the mountains?' "

Andy had been to Bermuda a dozen and a half times, and it was to be the first stop on his winter cruise. Robert was eager for the trip. So was a third crew member, Ginna Campbell, wife of Daniel, the young ocean crosser who was then over at Mugwump, pounding giant bronze bolts into the deck beams.

It was getting to be late in the fall, when weather and sea tend to be unpredictable. Andy knew they had better get going. He figured from satellite and weather reports that they had a small but perfect window to push out over the Gulf Stream and cruise into Bermuda comfortably. Friends had thrown a party for them the night before. Robert was thoroughly hung over and so was miserable from the start.

But things got worse as soon as they rounded Gay Head. The wind started blowing, and Robert didn't think about his hangover anymore. The blow seemed a good thing at first. A good wind was what they'd expected would get them all the way to the Gulf Stream and across in good time. But the wind kept getting stronger. For the first couple of days it blew hard off and on. Ginna got sick immediately, so it was just

Robert and Andy doing the sailing. Because Andy's automatic steering device wasn't reliable in rough weather, one of them had to be on the tiller at all times.

Robert would sleep when he could, maybe thirty minutes at a stretch, but whenever he was feeling awake, he'd be at the tiller so Andy could rest. They couldn't lose Andy. "I was expendable," Robert says.

The weather got bad when they entered the Gulf Stream. Robert saw the water change suddenly, turning black. They'd been riding swells, but the Gulf Stream was choppy and mixed. The sky was black, too. It was like hitting a wall. Robert's only thought was that he wanted to get across this thing as fast as possible. They were completely reefed down, and they ran the engine some, Andy telling Robert that as soon as they got across the stream they'd have a nice sail on down. Robert repeated the words in his head and aloud till they became nearly a mantra.

When they made it across, at the end of Robert's six-hour watch, the seas were still 15 feet—he could see the horizon only when the boat rode to the crest of a wave—the wind was blowing 30 or 40 knots, and Robert felt the sky was pressing down right on top of him, that scary black sky at noon.

It was then that Andy decided to put out a storm anchor, a big conical dragging device called a drogue, and hove to till the storm went away. They sat down in the cockpit and talked. They had no idea where this weather was coming from, how long it might last, or how bad it would get. They just had to sit tight and hang on, let the waves toss them but keep the boat from turning beam to, a situation in which they could be knocked down.

The wind picked up to 60 knots, and the seas rose to 25 feet—getting to be the kind of weather that isn't good at all to be in in a small boat. The three crew, strapped into harnesses, pumped as fast as they could. One of them was always above deck, on watch; the other two would try to rest then, maybe eat a peanut butter sandwich if they had the energy to try to get the stuff on bread. Water was crashing over the stern so hard it squirted through the closed companionway hatch. Robert stood on the ladder to try to seal it, first opening it all the way—and a wave came down so hard it put him on his back, staring up at deck beams. That wave knocked out the radio as well, so they were without communication.

Below deck, Andy's boat is traditional, spare but warm and comfortable, the mahogany woodwork varnished but not to a high gloss. All very practical. The cushions on the settee are forest green. Andy installed a coal-burning stove in the forward end of this main cabin, and there are lamps and an oil lantern in the galley. An old nautical clock is fastened to the bulkhead above a settee. Books are strapped into bookshelves. It's a cozy little home when it isn't filled up with salt water and being shaken like a paint can, as it was during that storm.

Andy got the radio working, and they learned that the storm had come out of Nova Scotia and would continue to blow hard from the northwest. They heard other vessels in distress; one, still in the Gulf Stream, sent out a Mayday. The crew on *Harmony* could only hang on and pray that their boat would stay afloat.

As the storm wore on for the entire day and then all night long, Robert felt fear in a way he'd never known it before. It rose up like something hard inside him, something solid. It was astonishing to him how physical and tangible fear was. After two days of this weather, he was exhausted. Every time a wave crashed into the boat, he wondered how much more she could take—would the next wave take her down, break her apart? They'd already had one near knockdown; another could end them. Fear rose in his throat, and he had to swallow it down. He had been in life-threatening situations before, but never for this long. One day and night, then two, then three. Scarcely eating or sleeping, losing all sense of time—losing much of his senses, period. It required a kind of miracle of acrobatics even to use the head or get peanut butter on bread, let alone eat it. And it was in such situations, he knew, that accidents were most likely to happen—when people got sick, didn't eat, couldn't sleep for days as a boat was tossed and slammed on waves. Robert couldn't believe it. He'd never been so wet and tired for so long. He had plenty of time to think about his choices, though; there was nothing else to do but listen to the boat crashing about on those waves. By then he was seriously wishing to God he was back on a ranch. That was good work, making sure the animals found enough food and found their way to water, caring for them—between ten and fifteen thousand of them on any given ranch. Being on a horse all day long seven days a week, sleeping at night in the tent you carried with you. Same as buckaroos had been doing for a hundred years. They made all their tools, the spurs, the bridles, the saddles. When a cow died—as cows did sometimes in winter from the stress of the weather—you skinned it and cut the hide into a big circle. Then you'd use a razor blade to cut this circle into a single long strip about 2 inches wide, then cut that in half down its entire length and braid the pieces into rope. From this rawhide rope, Robert would then fashion a 100-foot reata, a traditional Spanish lariat for roping bulls and horses. He'd made his reins and blankets out of horsehair.

Robert had loved Nevada best—that was truly the big country. Horses ran loose in the desert. You could be in another century; you could be in any century at all. He'd loved the rhythm of living off this land. He and the others at his camp would wake at four in the morning. They'd pulled their own chuck wagon with them, and the cook would have breakfast ready: sourdough pancakes, biscuits, sometimes steak and eggs. Along with the cook, there'd been a horse jingler who took care of all the horses. The jingler would bring the horses in by four-thirty—"jingling the *caveata*," it was called—anywhere between sixty and a hundred head for five or ten buckaroos, and the buckaroos would catch a couple of horses, saddle up, and ride off. They'd ride the horses hard, for twenty to forty miles, and Robert would sometimes go through three horses in a day.

He'd loved the desert and its solitude. He'd loved the traditional values and self-reliance of the life of a buckaroo. He'd loved making things by hand and living immediately, on the surface of the *earth*.

Andy remained calm throughout the storm, tending the boat, taking care of himself, looking out for the others, making good decisions about what needed to be done. He was concerned but steady. No one was going to say they were sure to make it through this storm, and his calmness in the face of that uncertainty was a kind of gift. If they made it, it would be Andy's intelligence and experience that Robert would thank for it.

On the third day the wind decreased. The water remained rough, but the wind let up enough that they could haul in the sea anchor and put up a scrap of sail and move. On the following day they sighted

Bermuda, and Ginna was able to make pancakes for everyone, the first actual meal they'd had since leaving. They were able to set their clothes on deck to dry. Landfall in St. George Harbor that night was exhilarating. It was dark, the passage into the harbor was narrow, and exhaustion made the work tricky. They set the anchor, made all lines fast. Their friend Phil had given them some big jugs of beer from a good microbrewery on the Vineyard, and they sat out and drank them on the now-still boat. And it seemed then that there were no greater gifts available to mankind than a jug of Off-Shore Ale in St. George Harbor, Bermuda, and being safe and secure and dry in this blessed boat.

When Andy returned from his half year at sea, cruising, he would remember that first leg of the journey well enough to comment on it: "It got a little lumpy for a while," he says.

In either case—whether it was a lesson in exhaustion and wetness and the physical impact of fear, or simply "a little lumpy for a while"—this journey might also be described as workmanship of risk.

III

Ross said on that first day of caulking, "This is when it really pays off to plank the boat upside down." Instead of lying on their backs and hammering straight up, they could chink and thunk away upright at eye level, either standing on staging or on their hands and knees in the flat aft section. Planking the boat right side up would likewise have been difficult, as would sanding the hull, which would in addition have been extraordinarily uncomfortable. Every step so far has been made immeasurably easier for having planked the boat upside down. The only drawback is that, having created a 10,000-pound object very close to the length and width of the shop, they now have to roll it back over.

"And that's fun!" Ross says, grinning. He might well have built the boat upside down even if there hadn't been any benefits to it at all, simply for the pleasure of rolling her back over.

A week and a half pass—during which time the seams are painted and then puttied, and then the hull is sanded once again and painted with enamel—before the day of the roll arrives. This gives Ross time to think it through. "Easing it over in your mind takes a long time," he says. On the day of the roll he stops by Bargain Acres to pick up some tools and pulls into the shop at eight-thirty. He walks to the outer shop and stares at the hull, silent for half a minute. Then he says, "That's a heavy object to roll over in a small space." He's still figuring out in his head how it's going to happen. The object is 32 feet long, only a few feet shorter than the outer shop; it's 11 feet wide at its sheer, which is now near the floor, and about 8 feet tall at the bow. On one side is a workbench built into the western wall of the building, and on the other, a beautiful wooden sailboat nearing completion.

Nat is already at the yard. "Most people," he says, "to roll that boat, would wait for a crane. It's amazing what you can do with a few simple tools."

Planks, rope, pulleys, a small hydraulic jack, and three warm bodies are what's needed. Typically, rollers—they lie around in the sand here, big round wooden posts, half the width of telephone poles and 3 to 5 feet long—would be included on that list of tools for moving heavy objects. Something that rolls is critical to the equation: you move a big object vertically with jacks and pulleys, laterally on rollers. Boats, whole houses, are easily pushed or pulled along if they're on rollers—yet another curve G&B relies on. But this boat isn't going anywhere; she's staying right where she is. So a different kind of roller is required, and Ross, after turning it over in his mind for a while, has decided to build the roller, the circular curve, onto the boat herself, like a frame. He will build two 90-degree curves—two separate quarter circles—out of sturdy planks and bolt them to the keel and sheer so they'll sweep over the boat's port side. He figures the radius of this circle, 6 feet 7 inches, by taking some measurements straight off the lofting-floor drawing. He and Nat and others build the circle fast, just as they would construct double-sawn frames, first making patterns, then sawing the pieces out of fresh white 2-by-10 planks and fastening them together using drywall screws and battery-powered screw guns. A half hour later they've constructed more than 20 feet of curve.

As Ross and Nat begin bolting the circles into the keel, Bernie Holzer, a purser on one of the Steamship ferries, a friend of the boatyard, and a close friend of Ross's, arrives. Bernie has worked on the water his whole life and visited most of the world's deepwater ports. He's got an unforgettably high-pitched, scratchy voice that sounds out in the shop: "Are you sure this is the way primitive man did it?" Bernie has come to watch, and to help if need be. "Well, I got twelve hundred and fifty Egyptians comin' over on the next boat!"

"Great," says Ross. "Let's stop for coffee."

Ross steps off the boat and resumes his staring and figuring—he's all concentration now—wondering if they can get by with just the one jack to lift the boat off her blocks and inch her toward the Bella, so she'll have room to roll.

"What about the hydraulic jack at your house?" Bernie asks.

"Unfortunately, it's holding up my house," Ross says. (He lifted it last weekend to put a foundation in.)

Ross is planning to roll the boat toward the workbench. He looks over at the Bella and explains, "If we roll it this way, we'll destroy two new boats." He knows that when the keel passes the 45-degree mark, *Elisa*—all 10,000 pounds of her—is going to want to right herself, and the bench will act as a stop if she gets out of control. But in order to give her enough room to roll—she's right up against the bench now—they must first move her closer to the Bella by about five feet.

Nat is on top of the boat, continuing to bolt the wheels to the keel, and Ross asks, "You mind if I start moving this over?" Nat shakes his head, and Ross and Bernie crawl under the boat. Ross begins the slow process by positioning the jack so that it's tilted away from the Bella; when the boat gets high enough, the tilted jack becomes unstable and rights itself, tipping back up straight and shifting *Elisa* several inches with it. In this manner they creep her across the lofting floor, Duane helping to move the supporting blocking as the boat moves, Bernie placing a steel plate beneath the jack to keep it from popping through the lofting floor from the weight of the boat.

With one lift and tip, the boat shifts more than Ross is expecting, and the entire building trembles. He says, "That wasn't supposed to happen."

The boat's owner, Jonathan Edwards, arrives and sees *Elisa Lee* for the first time. He grins, standing in the doorway, tall and lanky, balding, with close-cropped hair turning gray, cool blue eyes, dressed in a warm-up suit and running shoes. Nat sees him and says, "Heeeey!" and they shake. Ross calls out greetings from underneath the boat. Jonathan has returned for the spring and summer to help work on the boat in between concert dates. "It's beautiful," he says.

When the boat is in place—Ross just eyeballs it—and Nat has gotten the two wheels securely fastened to the port side, the process of rolling her can begin. The idea is to lift her up using chain falls, ratcheted metal pulleys with several purchases, which are attached to the rafters overhead and also to the fabricated boat wheels where they're attached to the keel. Nat will also run fat white rope from where these chain falls are, over the center of the boat, through the eaves, and down over the Bella, then fix it to one of the rails in the sand via a come-along, a kind of portable winch; this rope will be attached to the keel and will keep the boat from rolling too fast. If they lose control of her, all that weight now with wheels bolted on, someone could be seriously hurt, and the boat could be damaged, too.

Duane climbs on top of the hull and begins pulling the chains, lifting the boat by fractions with each long pull, the chain falling with a rattle against the planks, and the boat quivers first and the floor shudders as she just budges.

A spectator who has been watching this process whispers, "This is *insane*."

There are spectators! People have begun to gather to watch Nat and Ross and the crew roll this big fat boat inside a small shop. Gretchen has appeared at the top landing of her loft to observe; Simmy, Bernie's wife, has joined her. All sorts of friends and strangers have somehow materialized. Photographers from both island newspapers are here.

Listening to directives from Ross, who's on the ground watching and looking, stroking his beard, focusing hard, Duane pulls on the chain. The boat lurches forward, and the spectators take an involuntary quick breath. Ross, right in front of the boat, doesn't flinch—too focused. He tells Duane to keep going. And Duane does, but too aggressively this time, and the boat skids suddenly a foot toward the bench, and Ross shouts, "Duane! Duane! God—gosh darnit!" Duane manages to steady the boat by reversing the chain fall. Ross says, "You know you just took away about half what we did with the jacks?" A half hour spent shifting inch by inch, lost in a moment.

What makes the spectators' hearts pound as they watch this is not the fact that this creature moves fast; it's that they can tell from the rumble of the shop that it's going to move, but they don't know where or how much. And, too, it's that once it starts, it can't be stopped. No one can really see where the forces and the weight are. Everyone trusts Ross, though, and he seems literally to see the critical vectors that hold this whale still or press it to move.

And as *Elisa* moves, as he guides her up onto one edge, off her blocks, attached by rope to the building, her shape changes before your eyes. She begins to roll, inch by inch, and as she does, those elusive curves wave at you, undulate; her belly seems to swell like a squeezed water balloon. This optical illusion, combined with the breathless fact of the boat's potential to suddenly roll and crush someone or take down a wall, transfix the crowd.

The chain falls soon lift the boat off the blocks, and her heavier port side, where the wheels are attached, drops while the starboard rises. The boat is now irrevocably tilted, and Ross says, "It's started its journey." She looks, in a way, like a capsized boat, washed up on rocks, and Nat says to Jonathan, "You don't want to see it in this position again."

Jim says, "Eet looks like an elephant trap. We should get some peanuts." And he laughs.

Nat climbs up on the boat to work the chain falls, lifting the boat higher. It feels as if the entire building—and the ground, everything— were working to hold her up; the strain is like a hum in the air. The spectators are silent, the only noise the chain rattling on the hull, until the boat groans and shifts, sending another tremor through the boatshop.

Slowly Ross begins to pay out the fat white rope attached to the come-along attached to the rail, letting the keel turn gradually. He lets out a little more as Nat works the chain falls.

Ginny, taking pictures from her open office window, answers a ringing phone and calls out, "Ross?"

Nat says, "Ginny, no phone calls for a while."

"It's your niece calling from Cuba."

"Tell her to call back in one hour," Ross says, tending the come-along. "Explain what's happening and apologize."

The keel drops, and Ross shouts, "Whoa whoa whoa! Nat, you got it. I can see it wants to go."

She's crossed the line—she wants to be a boat.

Ross retrieves a long metal post that serves as a lever. The sheer doesn't touch the ground; the weight is all on the two wheels, the entire hull secured by various lines. Ross puts the lever under the sheer and lifts, and the boat scoots toward him with a rumble, and he says, "Whoa, whoa." Down to his purple long-johns shirt as the day has warmed, in his work pants and ratty tennis shoes, Ross appears tiny against the big, wide boat. It seems as if the boat could shoot out into him, right herself, and launch him into the rafters.

Jim, standing and watching from a distance, says, "I wouldn't want to play there."

Ross tries to figure where all the pressure points are and then says, "It's just going to scoot. It's never going to be out of control."

Nat climbs up into the rafters now, and the boat continues to roll until she's on her side, a quarter turned and completely vertical, resting against the bench. Nat climbs out of the rafters onto Gretchen's landing and descends the stairs, grinning. This is exciting. He knows exactly what to do. They'll wedge planks beneath the wheels, attach bar clamps to the planks and a come-along to the clamps and to the rail, and then simply pull the planks, and the boat will come with them. He instructs David Stimson to slather the bench with marine grease so the wheels will slide easily over it, and says, "I'll go get some ball bearings"—sand, that is, to throw under the planks to encourage some slide.

The planks are inserted using a lever and a sledgehammer; then they're attached to rope and come-along, and sand is scattered beneath them. Ross works the come-along, Nat and Bob Osleeb help to guide the boat, and inch by shop-rumbling inch, she turns, and turns, with increasing ease as the weight of the keel nears the ground, until Ross yells, "Yippee, there she is!" and the keel comes around completely and the boat is righted. Ross strides over to see where she's ended up and says, "She's almost at that center line where she began life."

Everyone is smiling. The entire process has taken more than eight hours. They've never performed such a maneuver before, and it's been a success.

Before they secure Elisa, get blocks under her, and make her perfectly level, Nat and Ross will take a moment to relax and revel in the sight of this new boat, in what is a transformed shop. Suddenly there's all sorts of room, and they can now see what she really looks like. They both walk back to the bench to lean on it and regard their work. In an unlikely bit of chance, each chooses one of the points on the wall-length workbench where the wheels ran their course, each rests against it at the exact same moment, and both shout "Ah!" simultaneously, as if the bench were electrified—mirror images, each having gotten the surprise of marine grease on hands and clothes. They look at each other and laugh. Different as they are, they are perfectly in sync.

IV

S pringtime comes late to this island, and everyone grumbles
about it. Foliage opens up down-island while up-island trees remain bare—the effects of the wind on the differing landscapes
are that marked. But the sun shines more brightly now, and later into
the evening after daylight saving time comes, and the freezing sleet of
late March ends, the earth begins to warm, and the island's fragrant
lilacs begin to bud.

Ross and Nat direct work over *Elisa Lee,* which is now looking
very much like a big old lobsterboat. The next order of business is to
saw out angelique sheer clamps—long planks that will run the lengths
of the upper and lower sheers, attached to the insides of the sawn
frames—and bilge stringers, which will likewise run the length of the
boat down in the bilge. Once the boat has been rolled upright and
these long interior pieces running fore and aft have been cut, planed,
and fastened, work on the pieces running athwartship becomes the
focus. Deck beams—three of them are still bolted to the landing outside the shop, bent over a jig—will be sawn out and will rest on
the sheer clamps. The engine bed must be bolted in, as must the final
floor timbers that will support the cabin and cockpit soles, and the
bulkheads—thick, heavy pieces of marine-grade plywood that will
form the walls of the boat, separating cockpit from cabin—have to be
fashioned and fastened. Ross thinks of bulkheads as big knees—they
join everything together and make the boat stiff and sound.

Nat constructs some of Bella's spars, building them hollow, out of
spruce. Painting and varnishing and general readying of the G&B fleet

begins over at Packer Park, where James Taylor's *Sally May,* John Evans's *Swallows and Amazons Forever,* David McCullough's *Rosalee,* and a dozen other G&B-built boats have sat on jackstands beneath tarps all winter. Buckaroo Robert will be at work here daily through spring, till the boats go back in the water. And so will Kerry Elkin, the former cranberry farmer, who has found he enjoys this work and intends to pursue it. David Stimson will join them in the sanding and painting of hulls.

Ross spends several April days away from the shop, at Bargain Acres, where he and Brad Ives are erecting a big pole barn to shelter Brad's wood.

The poles they're using are old pilings that once supported a retaining wall; they're saturated with creosote, encrusted with barnacles. They were given to Ross by the friend who took the bulwark down. These pilings are now in the ground and rise 14 feet. In a day or two the trusses from the old Ag Hall sheds will go up. The only materials they'll need to buy for this 28-by-56-foot barn are the fastenings and the corrugated roof. Brad has cut notches in the pilings to connect them with beams. Ross has built staging right onto the back of the G&B pickup truck so they can drive it around to where they need to work high up—it's a sort of movable ladder. Beams are fitted into the notches and bolted in along the entire rectangular perimeter. Brad is up on these poles, his feet dangling, taking measurements to determine how much adjusting they'll have to do to make all the corners square. The trusses are 28 feet at their base, rising gently to a peak, and weigh about 200 pounds apiece. Twenty of them, in all, must be raised 15 feet in the air, maneuvered onto the top of this growing structure, and bolted in. Ross rigs a pulley system over the poles to hoist the trusses on a block and tackle that's attached to a pulley; he'll roll the trusses, one after the other, along the cable across the front of the building to their final position.

Ross heads into the pines facing the pole barn to find a tree on which to rig the front of the pulley system for the trusses. Brad remains on top of the structure, still measuring.

"I can't believe it," he shouts to Ross. "We're square. We're off by one inch."

"One inch?" Ross says, invisible now in the woods.

Brad, all skinny height and angular features, his fine hair short and dark, climbs down and confers with Ross about the progress. The sun is low, and Ross says, "Well, should we call it a day?"

Brad looks at his watch, nods, and says, "In five minutes I've got to call my Amerindian friends in Suriname."

✳

Brad is an integral part of Gannon & Benjamin, but at the same time he skirts its periphery. Sometimes he's working for G&B, and then you look up and he's gone. If he's ever in a group (and this is not a common occurrence), his height and hollow, angular features make him the focal point; his quiet, certain voice commands attention. His is a spiritual presence, at least among the builders here—he's the primitive sorcerer/magician first identified by Ted Okie. True, Brad Ives on the surface appears to be an unremarkable middle-aged man: upper-middle-class boy of the 1950s, recent home owner with his fiancée, April, soy-food-products salesperson, and the only G&B boatwright with a cell phone, a laptop, and a business of his own, Deep Water Ventures. But it's also true that while he was born into conventional Eisenhower America, he's spent most of his adult life on the shifting, lawless-but-for-nature, fluid surface of earth.

Ginny Jones, who is about as skeptical as they come, a self-described curmudgeon not prone to delicacy or gush, says, "Without that wood connection, we'd be screwed." Then she adds, "Brad's an incredible resource and an incredible person."

Springtime for Brad means another trip to South America. Then again, so does fall. And sometimes summer. He tries to avoid May through August and early winter, when the rain prevents the cutters from getting logs out of the forest, but for him, Martha's Vineyard to Suriname, a Third World country almost completely devoid of the influence of American mass culture, is becoming an ordinary commute. He is so comfortable in Suriname that his life there, as he recounts it, seems almost American-suburban. Last August, returning to his business and restoring old contacts in his efforts to buy timbers for *Rebecca,* he wrote Ginny a leisurely e-mail that began:

I have already been to one of the mills today to mark logs for cutting to length as they came off the barge, but all was in an uproar as the two brother owners had been called by the watchman at 0630 that some guys were stealing wood. They went down and followed the wood just up the road. Radjen waited outside while Lloyd went to get the cops, but they wouldn't come. Too lazy or disinterested as thieving goes on all the time in that part of town. I worked most of yesterday as well, painting ends and measuring. The work is going OK, better than last time. Now I think they take me a little more seriously as I keep coming back. Maybe the next time they will even store some logs for me. The silverballi logs are smaller this time, but with some shape, so I try and match up the tree to the boat. Better use of the tree and more satisfying to me.

Home life is kind of disastrous again. Jan, my Dutch roommate, is all busted up from a bad bike fall, and Rosita has been sick with a kidney infection. I am nursing them and doing the cooking and cleaning up. Also there is a lot of gardening work that needs to be done as our yard is a jungle. I kind of like it, as the weeds are like small trees, some with pretty flowers. The yard is full of overgrown leggy impatiens as well.

In another e-mail, he wrote:

In the evening I biked home to my Hindustani village, found the house filled with women from Miti Switi, the internet dating service run by my housemate, Jan, and then biked a few kilometers to a sleepy little Javanese village where Bongo Charlie lives. We had a great dinner and short jam session on the many homemade percussion instruments Charlie makes, until riding home in the midnight hours enjoying a fine starry night off from the rainy season.

Brad has cleverly managed to combine the components of this earth that he loves best—boats, wood, boatbuilders, and the Third World—in his work, which is, as he says, simply this: matching up the tree to the boat.

His trips have grown more frequent since the *Rebecca* project returned him to the country where he found such plentiful wood in the 1980s, before civil strife and political murders made life there unpleasant, and before a short rest of his own in federal prison, and the ensuing years of probation, kept him landlocked in the United States. Wooden vessels big enough to require his rarefied work near the equator continue to be built, and his orders are increasing. Having departed from his sister's home in Boston, he arrives in Suriname with his biggest order ever, for 50,000 board feet of angelique, wana, and silverballi that will be trucked to yards throughout the United States and then sawn, planed, chiseled, and bent into big boats: *Gazela of Philadelphia,* a 170-foot barkentine being restored by the Philadelphia Ship Preservation Guild; *Lynx,* a new 80-foot schooner being built at Rockport Marine, the yard that launched the last W-class cold-molded racing sloop; a Malabar Sr., a 59-foot schooner, in upstate New York, and a similar schooner in Seattle; as well as future G&B boats.

✳

The old DC-9 out of Curaçao touches down in Suriname five hours late, and Brad, dressed in a short-sleeved shirt, jeans, and old running shoes, waits in line to show his passport and visa and then stands in the dingy yellow cement-walled baggage area, where a single conveyor belt runs. It's after one in the morning. The air conditioners struggle in the tropical humidity. He leans against a square pillar with one bag at his feet, along with a scuffed black briefcase containing his traveling papers, wood orders, boat drawings, his glasses, a Robertson Davies novel, *The Fifth Business,* and $9,000 in American cash. After forty-five minutes, the last of his baggage arrives.

He walks past the Customs agents, who don't delay him long, and into a mob of taxi drivers grabbing at his wrists. Jan Blom, a tall, slender Dutchman with short, blond hair and blue eyes, calls, "Brad!" and rushes to help him with his bags.

As they walk to the car, Brad asks, "Have you been waiting all this time?"

"Yeah," grumbles Jan, who's spent more than six hours in the parking lot, not knowing if any plane at all would come in.

The airport is less than an hour's drive from the capital city of Paramaribo, along a paved two-lane road sided by fragrant tropical brush. Dense, cool air rushes through the open windows as Jan briefs Brad on the latest news. Bongo Charlie is in jail awaiting trial for possessing three grams of pot; their friend Naseer is in the hospital with back problems, unable to afford the $700 flight to Holland and free medical treatment. Teachers are striking for decent pay, so school hasn't started (half of last year was missed, too, because the government failed to make the payroll). The building of two bridges spanning the Suriname River, though—an "improvement" that will cost $200 million and that according to many is unnecessary—remains on schedule: payroll has apparently never been a problem for President Wijdenbosch's pet project. Meanwhile, Jan's Internet dating service, designed to pair Surinamese women with prospective husbands from Europe and America, is not the booming success he'd hoped for, and he'll soon be broke.

Brad leans his head all the way back against the headrest and lets the early-morning air wash over him. He's tired, and he's got the scratchy beginnings of a cold in his throat, having risen a little before four the previous morning at his sister's Boston apartment to begin this trip.

Jan pulls around a small cement divider leading to the line to drive onto the ferry. Stray dogs roam the crumbling streets of this once-prosperous Dutch colony, now composed of Africans, Amerindians, Indians, Javanese, Creoles, and Dutchmen. Jan's is the only car here at the ferry landing on the edge of a large, deserted, unlit square in town, though a couple of shadowy figures loiter here and there, and a taxi driver fixes a tire beneath a distant streetlight. Brad and Jan must wait another forty-five minutes to cross the mud-brown Suriname River, and they get out of the car to stretch. Brad takes a drink of water from a small bottle he bought in Miami, then leans against the car. A vanful of French Guianans, having loaded their vehicle with vegetables and goods to take across the river and back to their neighboring country, pull up beside them, and everyone gets out to reorganize the packing of their goods: bundles of limp, foot-long green beans called *kosse band;* cassava, a root vegetable that will be grated and stuffed into a long woven tube called a *matapi* (which they also appear to have purchased) to allow the poison to drip out and the cassava to dry, after which it will be cooked on a hot plate into hard cakes; and various kinds of seafood and other vegetables. The oldest person from the van, a black woman wrapped in colorful cloth, her hair pulled back tightly, looks the two lanky white men up and down with amused skepticism, hands on her hips. Tilting her head back and gazing at them over high cheek-bones, she speaks in Sranan Tongo, the lingua franca here. This is a cre-ole language, not a pidgin, and it's easy to get the hang of. A boat club down the river, for instance, is called Watramamma, or "the Mer-maid." Sranan Tongo means "Suriname Tongue" and is sometimes also called Taki Taki. Jan manages a few words in reply. The old woman cackles and saunters off.

The ferry, an old metal barge with a rail around it and a large pas-senger cabin, arrives late for this last leg of Brad's journey. By then a line of cars and trucks has formed behind Jan and the French Guianans. The ride itself lasts less than ten minutes, at the end of which time Jan restarts the engine, waits to disembark, accelerates off the ferry ramp, and speeds down the four miles of paved roads to his termite-infested, wood-and-cement, two-story dwelling. Jan and Brad remove their shoes before entering the kitchen. Brad turns on the kitchen faucet, but there's no water at this hour. The roosters in the farmer's yard across the street begin their incessant crowing as Brad hooks mosquito netting to the ceiling above his bed in a bare upstairs room. Soon the noisy minibuses filled with people headed for the ferry will begin their routes, followed by motorbikes, engines buzzing at 6,000 rpm. More than twenty-four hours after waking in Boston, Brad sleeps—but only for a few hours. He's eager to get to the mill to see what raw logs await him.

✴

The following morning, toting his briefcase and wearing a small backpack, Brad walks out onto a dock and boards one of the couple of dozen passenger outboards that make the half-mile crossing all day long. The wana-planked boats are shaped like fat dugout canoes, with low roofs and 25-horsepower outboards; they're 30 feet long and carry as many people.

Lloyd Baldew greets him in an air-conditioned back office of the Toeval Mill. Lloyd, whom Brad describes as "the bull," is a manic, aggressive businessman of Indian descent, forty-one years old, with short, dark hair, dark eyes, and boundless energy. Lloyd directs Brad's attention to the new unfinished flooring after greetings have been ex-changed. Much of Toeval's wood winds up as hardwood floors; when Ross lays the floor of his house, it will be with wood from this mill.

"That's red locust," Brad says, regarding the new floorboards. "There's angelique. What's that?"

"Guess, guess!" says Lloyd, eager to show off the new white hard-wood they're trying to introduce in Holland and Belgium. Brad cannot guess.

"Riemhout!" Lloyd says with proud satisfaction.

They sit to talk about wood as coffee, served in glasses, is brought in. The walls are plain, and because Lloyd is in the process of redoing this room, there is little furniture here other than a desk, a table, a few chairs, and a refrigerator.

"We have nice logs here now," says Lloyd, in his excitable manner. He's been preparing for Brad's arrival. "Fifty of the big logs. But no wana. How is Brazilian wood in the States?"

"Big, but not their main market," Brad answers. "Europe is their main market. Last wood I brought from there was four or five months late."

Brad has brought his laptop, and Lloyd, who thinks of him almost as family (he and his brother, Radjen, are invited to Brad's wedding), makes him feel welcome to set up camp here. It's indicative of the way Brad works in the Third World: he knows how important it is here to make and maintain friendships, and he also seems genuinely to like the brothers.

Brad removes a fat white envelope from his briefcase and pushes it across the desk. "I want to pay a deposit of six thousand dollars for my wood now." He will deal only in cash while he's here, unwilling to use local banks until he can assess the political and economic climates. The country's infrastructure is no longer dependable.

Baldew discusses what he knows is now available in terms of logs, noting that Rudy Rageomar, the mill's sixty-eight-year-old foreman, is expecting Brad. "He's already sawing for you," Lloyd says.

Brad says, "I'd better go out there now." He doesn't want Rudy sawing before he's seen the logs.

✸

Brad's first and most difficult order of business is a rush for *Gazela:* the waterways, a section of the boat similar to a covering board that runs the length of the sheer between the frames and deck. They'll need some sweep, if he can find it. The pieces that Brad will saw will be nearly 1 foot square and 20 feet or longer, wood that must be nearly perfect, with straight, clear grain and no major faults. Put end to end, these fat pieces would run well beyond the length of a football field.

Brad strides into the 90-degree tropical heat to see what the Baldew brothers have for him. The partially enclosed section of the mill is about 300 feet long, with one wall and a corrugated metal roof supported by I-beams. Some wood is finished and neatly stacked. Brad walks between mountains of scraps, the ground a spongy mixture of sawdust and dirt. Beside and beyond the mill, strewn about in the grass, are the enormous raw logs of angelique and silverballi, each 3 or 4 feet thick and cut in half to a length of between 30 and 40 feet.

Brad never moves quickly; he is slow and steady throughout his twelve-hour days here, as if to conserve energy in the sultry air. What he sees concerns him. The angelique pieces are not as big as they might be. He climbs over the piles of enormous logs—a giant's Lincoln Logs—measuring as he goes, scrutinizing the shape and the quality of the ends, trying to imagine the interiors. In order to yield a good piece of wood 1 foot square, a tree must be well over 30 inches in diameter, to allow for an inch or two of sapwood on either side and exclude the heart, which typically contains faults and drying cracks. But Brad finds few logs that large.

"Look, Mr. Brad!" Rudy Rageomar calls. "I have a nice piece. I cut it."

"It's not big enough," Brad says, measuring.

He's got a potential problem, and he heads with Baas Rudy back to the cool office to call John Brady, who's in charge of the *Gazela* project.

"John, I'm down in Suriname, and I've got some questions for you," Brad says. "Those twelve-by-nine-and-a-halfs, can those be boxed heart? . . . Is that all right? . . . Can you see the ends of them? . . . Well, we can, but it will take a long time. . . . It'll be better, in fact, because you won't be getting into the heart crack at all. . . .

And the samson posts? . . . You want me to saw out to twelve? . . . You got it. . . . Yeah, I got a sore throat. I'm all right. I got in at five in the morning. Took twenty-four hours to get here. . . . All right."

Brad turns to Rageomar. "OK," he says, "that's no problem with the heart." Brad will take the enormous pieces straight from the center of the log. "But we're going to change a little bit how you're sawing. Do you have your list?" Rudy nods and removes a sheet of notebook paper from his shirt pocket. "We're going to make that nine and a half," Brad says, and he continues through the order with Rageomar, who will preside over all the sawing. The finished pieces in the boat are to measure 9 by 9 inches, and because the heart crack will be in the center of each piece, not on the surface, he can cut the logs more precisely: he'll leave the 12-inch width so that the *Gazela* crew can cut some sweep in as needed.

"You see," Brad says softly to himself, exhausted from the journey and sick with a cold, "this is why you've got to be here." Each tree, each piece of wood, is unique and therefore requires its own strategy for cutting. Moreover, he often has the exact designs and drawings for a boat with him, so he can say, as he did a year ago, This log, cut this way, will give us the keel for *Elisa Lee* as Nat has drawn it here, or, This piece will give us a possible garboard for *Lynx*. This method is extremely time-consuming for Brad, but it reduces waste and expense once the wood arrives at a boatyard.

"It's amazing how complicated cutting wood is," he says, heading back out into the heat, which by eleven feels like an actual weight. "It's an art." By the time the order is done, he will have graded, inventoried, bundled, and loaded hundreds of pieces of wood into Fort Lauderdale–bound containers, each piece cut for a specific part of a specific boat, predestined at this mill to be a plank, a keel, a deck beam, a waterways, a sheerstrake.

"It's often frustrating and heartbreaking to see how much of a tree is wasted," he once explained to Ginny in an e-mail, his main form of communication with America when he's here. "The outside sapwood is not durable, and the heart is almost always full of cracks and soft spots, and there are other defects running throughout the tree. But we do our best and sometimes the planks are beautiful one after another. From the tree to the ship is a wonderful puzzle, and every piece is different. I am still learning plenty about the different subspecies and how they grow and how best to saw them."

Brad chooses the logs, and Baas Rudy directs a log loader to bring them to the horizontal band saw, built on rails and designed as a portable mill that can be taken into the forest. Today the first of an eventual 100 cubic meters, in lengths up to 40 feet, pass through it.

A motorized block and tackle lifts the sawn pieces to an upper level and the gang saw, a big, muscular construction of iron, originally built in Germany, that would feel right at home in the woods outside Ross's house. Numerous vertical blades can be fitted into this saw, so that a big timber 2 feet square can be driven through to become seven or eight 2½-inch planks—"like a pasta maker," Brad says. Today and tomorrow, for the waterways, two central blades will be set up 9½ inches apart, and additional blades will be set on either side of them at a distance of 2½ inches. A 24-by-24-inch piece 20 feet long will thus result in one waterways for *Gazela* and several planks for *Lynx*. Brad will inspect each piece for cracks, knots, and imperfections that might weaken it; note its dimensions; and number it for his inventory.

The kind of wood Brad imports—cut individually to these specified lengths and these dimensions, and of this quality—is difficult to find in America (Brad charges between $2.80 and $3.50 a board foot for it, a reasonable price). Moreover, he has created this business at a time when diminishing supplies of Honduran mahogany and Burma teak are growing increasingly expensive, and the erosion of wood resources worldwide is evoking serious concern in terms of its effect on the environment; thus his work of measuring and cutting each individual piece sawn from handpicked trees raises important ecological issues that extend beyond the work of the boatbuilder—issues he's very much aware of.

✳

Brad has been at the mill for only a day when he tells Lloyd that he needs bigger logs than he sees here. So the following afternoon, after sawing more raw logs, he hops into Lloyd's white Toyota 4Runner, and they head sixty miles east into the Baldews' concession in the dense Surinamese rain forest.

Since Brad first arrived here, in the early 1980s, several groups have formed to promote proper management of the earth's forests. Perhaps the most respected is the Forest Stewardship Council (FSC), headquartered in Oaxaca, Mexico. Established in 1993, this nonprofit group certifies forests throughout the world as being properly managed, and labels products originating therein to assure buyers that the wood comes from forests managed under internationally recognized standards of infinite sustainability. Britain is the most active market in this sector; about a quarter of all the wood sold there is certified. In all, some forty million acres are certified, in thirty-one countries, numbers that double yearly. Home Depot, the largest buyer and seller of forest products in the world, announced in 1999 that it would begin giving preference to certified wood.

Suriname, a little smaller than Missouri in total area, is composed mainly of forest, one of the most diverse ecosystems in the world, and until recently suffered little deforestation. Lately, though, the Suriname forest has been threatened by the sale of large tracts to foreign concerns—concerns that last year, according to one official (who has requested anonymity for fear of losing his job), exported 20,000 cubic meters of raw logs, not only thinning the forest but denying the citizens of Suriname potential income. "The rain forest is big," the official tells me, "but if they do this for years, it will be dangerous."

Brad and Lloyd cross the Cottica River at Moengo, where Toeval logs are loaded onto barges and floated to the mills. The car rolls over muddy roads colored a deep red from bauxite, the mineral used to make aluminum and the country's main source of income, toward the first of three Bush Negro camps on the Baldews' concession, which totals 108,000 acres in all.

The mill employs about a dozen Bush Negroes, descendants of Africans enslaved by the Dutch in the eighteenth and nineteenth centuries, who now live throughout the rain forest. Toeval employees occupy small camps in rain-forest clearings and locate the dozen species milled by Toeval by hiking through the forest, wearing little more than towel-like garments wrapped around their waists, chain saws in hand. They take each tree down and drag it out of the forest with a backhoe-sized truck called a skidder. Log loaders then lift the trees onto an ancient Mack truck rig that transports them to the riverside landing at Moengo, beside an abandoned American village, a kind of presidio of

barracks, stores, a school, all built by Alcoa for the families of employees who once worked here mining bauxite. At Moengo the logs
are loaded onto a barge, 200 tons at a time, and a little homemade tug
with a 90-horsepower engine pushes the barge down the Cottica
River to the Commewijne River and up the Suriname River to the
mill—a twenty-five-hour journey if nothing breaks down and all goes
perfectly—where the logs are off-loaded and strewn about in the grass
and dirt until they're needed.

When Lloyd and Brad arrive at the first camp, they find Willem,
Djanie, and Ahmed beneath their shelter, a large blue plastic tarp, with
dirty pans stacked on a log, a fire smoldering, the youngest holding a
toothbrush heaped with toothpaste. Today was unproductive, the lead
cutter explains to Lloyd, speaking in Sranan Tongo, because the skidder got stuck in the mud and it took them all day to get it out.

Lloyd is angry that his cutters haven't found bigger logs, and he
says, "I told you last week Brad was coming. He only buys big logs. If
he doesn't buy them from us, he'll buy them from someone else.
Where are the logs?" He threatens the most senior of the cutters: "If
you don't get the big logs, I'll take the skidder and put it somewhere
else." (This would leave them with three chain saws and immovable
logs for which they wouldn't make money. Lloyd has to put some pressure on them, he feels, or they won't find Brad's logs.)

Brad stands behind Lloyd as Lloyd speaks. A large fern reaches up
at his leg. He stares at it for a time, then strokes its leaves. The leaves
clench instantly around their stem—a visual reminder of how alive this
place is, how sensitive. After more discussion (Brad holds his hands far
apart to show how wide he needs the logs to be), he and Lloyd leave
for another camp, along rutted roads, narrow alleys through the forest,
stopping where logs have been dragged out to examine them.

In the car, Brad speaks for the first time of his long-term plan to
bring his former first mate Jack Risser—an "educated econut," in his
words—down here to explore ways of regenerating the forest, or at the
very least of propagating the species Brad's taking out.

"We're getting out the big ones and letting the small ones grow,"
Lloyd says impatiently.

Brad gently tries to explain that he hopes to do more than that,
wants actively to regenerate the forest. "There would be no cost to

you," he assures Lloyd. "I would like to make sure that the forest is not being degraded. My business is big enough now that I can pay Jack to come here."

Brad can't yet practice certified forestry management by himself. Nor is he obliged to: his entire operation is legal according to both the laws of Suriname and import laws enforced by the United States. Furthermore, the amount of wood he takes out of the forest—this year about 500 cubic meters, on two hundred trees—has a negligible impact on the rain forest: "What Brad is doing is nothing," says the publicity-shy Surinamese official, who's familiar with Brad's business. In any event, Brad wouldn't sell any more certified wood than he does uncertified wood now, so insisting on forest management doesn't benefit him or his business in the least; on the contrary, it's a lot of work for which he won't make any money, and he may even lose business because of it if he has to raise prices higher than builders are willing to pay. That's not his first concern, however.

He remains intent on ensuring that the Suriname forest is managed according to internationally recognized standards because it's the right thing to do—the forest provides his livelihood, after all, and he's been in countries that have obliterated their forest resources; he's witnessed the devastation.

Richard Jagels, a professor of forest biology at the University of Maine and the wood columnist for *WoodenBoat,* is skeptical of any importer who brokers wood from forests that have not been certified by the FSC. The Toeval Mill uses what's called selective cutting, but Jagels maintains that the term doesn't mean anything. "They're 'selectively cutting' the most valuable trees," he says. "They're mining the forest, not managing it." But he adds, "The one thing you can say for boat-builders is that they use such a small quantity of wood, it's almost irrelevant."

Scott Landis, a writer and editor active in forest-management issues who edits *Understory,* the journal of the Certified Forest Products Council, for which Brad has written an article on the Suriname lumber industry, says of him, "I believe he's trying to do a good thing. But the larger issue is that it's just impossible to know what an independent claim [of stewardship] means, if anything at all. It's hard enough for the FSC." But even the least effort can raise a country's consciousness, he emphasizes, regarding the value of universal standards in forestry management. If Brad can be one force leading the small country of Suriname in that direction, then, the impact of his business will extend beyond the boatbuilders to whom he is devoted, back to the source of their work, to the wood itself.

"We're not going to go right away for a stamp," Brad says of FSC certification, "because we don't want the government involved. . . . We will develop, and bear the cost of, our own pilot program with our suppliers, drawing from international research and documenting it well for our customers, who will pay for it with a surcharge on the timbers once the system is in place." (By the spring of the following year the Baldews will decide to seek FSC certification.)

"Suriname, being a small country, really has an opportunity to be an example of sustainable forestry—without even trying," he continues. "They already have it because of the size of their economy."

*

Lloyd drops Brad off at Jan's after eight, darkness having fallen while they were still out hunting logs. Jan, two Miti Switi women, and two children in their charge are finishing dinner at the kitchen table—stewed chicken, *kosse band,* rice—and Brad sits down to join them after his long day milling logs, grading cut pieces, and bouncing around the unpaved rutted roads of the Baldew rain-forest concession in Lloyd's SUV. The girls offer to do the washing up. There is no running water after dark, but someone has filled a large bucket by the sink for cleaning and drinking, and there's not too much junk in it. Earlier the electricity went out for twenty minutes, and no one said anything other than "Where are the candles?" After dinner Brad has a shower. There's a large bathroom area that holds a partially working refrigerator, a sink, a tile floor with a drain, and two other, smaller rooms. One of these contains the toilet, which was built to accommodate running water, but tank and pipes are not included here, so you flush by pouring a bucket of water into the bowl, which purges it, sort of. Beside this room is another, identical one featuring a spigot above a large plastic washtub filled with water in which floats a smaller bucket for dousing yourself. Brad will bathe, then check his mail on Jan's computer.

He'll be in bed early, in his bare room beneath the mosquito netting, thinking about tomorrow, when he'll have to visit a couple of other mills to see what they might have for him. He's got to hunt and hunt. In a few days, biking around in the Chinese yard, he'll spot a big angelique log, 40 inches in diameter and 32 feet long—a real keel log, large enough perhaps for a major part of the backbone of *Lynx,* an exciting find.

During his long days, he'll often take a break to walk down the dusty street to a small restaurant for *rotis,* crepelike flatbread filled with a chicken or lamb or vegetable curry. No utensils are provided; the *rotis* are eaten by hand, with a sink against the back wall available for washing before and after the meal. He's partial also to the cassava soup, the *heri-heri* (plantain, cassava, and vegetable stew), and *soto,* an Indonesian chicken noodle soup. He likes the unusual mix of Indonesian and Indian culinary traditions. Down the street from the mill is a sprawling market that offers all the cultural bounty of a country composed of six different peoples, where racial animosity is rare and racial violence is unheard-of. The people are friendly—gentle, even.

Brad is enormously comfortable in the Third World. At any moment of any day he can marvel at his surroundings and enjoy the drastic contrast between them and his own, overstuffed culture, happy to be away from the malls and the highways and the relentless pursuit of material comfort. "It feels good," he says. "I've lived a long time at sea and in Third World countries. You develop an appreciation for the immediacy and the economy of life."

I've accompanied Brad on this trip, and what I find most striking about Suriname is the absence of all things American. The government is so badly organized that it has created no tourism infrastructure. For instance, it has set aside 3.75 million acres of its rain forest in the belief that ecotourism is a growing business, but there are no roads leading to this extraordinary patch of the interior, or through it, for that matter, so you can't get there except by plane, and once you do, you're on your own. A *New York Times* article about the Suriname rain forest described it as "one of the least traveled places in the world"—a blessing that the government has been too inefficient to squander.

Because there is no tourism, there are few Americans, pockets stuffed with cash, wandering the city streets and demanding their American soft drinks and American fast foods and maps and postcards and T-shirts and Things to Do and Places to See. Paramaribo is that rarest of places, a developed city, a land, offering a glimpse of humanity untouched by American mass culture.

America's is a consumer culture, so the consumer determines how things run in the United States. Planes, for example, are pretty much on time, and anyone in the least inconvenienced by a delay screams bloody murder or stews with disgust and fury (witness the class-action suit recently filed against a major airline for an eight-hour runway delay). When a plane has to sit on a runway even for an hour, the flight crew announces its takeoff position like a Dow Jones update, and the attendants pass out earphones and show movies. When a flight is canceled, diligent workers at the ticket desk click away on their keypads to find all the customers other flights, often ones that leave earlier than the flight they were scheduled to take.

In Suriname convenience is not an issue.

Friends sometimes tag along with Brad to Suriname because they're curious; it's a place they would never otherwise go if they didn't know him. Mark LaPlume, an artist and boat worker who was part of the *Rebecca* crew, flew down to stay with him once, and just before he was to fly back to America, the whole country went on strike and no planes flew. He could've waited it out if he wanted to, but who knew how long it would last? Mark had to travel to a neighboring country and fly out of there.

The day I am to depart on an 11:00 A.M. flight, I arrive to find the airport deserted but for someone pushing a broom across the cement floor and two young men leaning back, drinking tea, and smoking behind a ticket desk. "No plane today, come back tomorrow at four A.M.," they tell me. I'll have to cross the river at two-thirty in the morning, locate a taxi, drive through the darkness, and, well, hope I see a plane. As it happens, I will, but you never know. And either way it doesn't matter, because that's not the operative fact, plane or no plane. The operative fact, I realize only after my incredulity at the deserted airport passes, is that I'm in Suriname and not in America, and that's a good thing.

The haphazard nature of travel in a Third World country is a chief asset for Americans. Sometimes things run, sometimes they don't, and because you never know which it will be, you have no expectations.

This may be why Brad is so relaxed when he travels to Suriname. He brightens here. He laughs and smiles more easily. A weight seems to lift off of him. He becomes a part of the place, even adopting a slight Sranan Tongo accent despite the fact that he doesn't speak Dutch (though he will speak in Spanish with a ticket agent in Miami and in French when he runs into a French mechanic at Toeval, and he's equally conversant in Portuguese).

When you have no expectations—nor the distractions of museums, ruins, and restaurants, T-shirts and postcards—then the sights and smells and nuances of the daily life of a place grow vivid. And for all its otherworldliness, its remoteness from an America where pretty subdivisions and malls roll out across the land like smothering Astroturf, Suriname feels humane and distinctive. You begin to sense, as Brad puts it, "the economy of life and the immediacy of life" lived here.

Brad has lived with this kind of immediacy for most of his years, sensed his own need for it even before he dropped out of Harvard to buy a wooden ship and sail it around the world, all the while steeping himself in many different cultures, mostly in the Third World or in the otherworld of the ocean's surface. Here, he has thrived.

✳

Brad Ives was born in 1949 in Providence, Rhode Island, the son of a lawyer who would eventually move to Washington to work in the Kennedy and Johnson administrations while young Brad remained at an East Coast boarding school. He was an independent boy, the "cooler older cousin" to other progeny within the family. Brad left Harvard not long after he arrived in 1968, but at the urging of his father, he later returned, though remaining restless. When a friend had the idea of buying a big boat and creating a floating commune to sail around the world, Brad was interested. So he and a small group of hippies, draft dodgers, and other disillusioned college students met in Vancouver to discuss the plan, find a suitable boat for sale, and pool their money. Brad went back to Harvard only long enough to persuade an instructor to teach him celestial navigation, a course the school had once offered. In 1969 the group traveled to Sweden to meet their ship.

Sofia, a 90-foot Baltic trader, sailed with a dozen crew through the Keil Canal in Germany, to Portsmouth, England, then south to Vigo,

Spain, and on to Portugal, where they set about rebuilding the old wooden cargo vessel and converting her to a three-masted schooner.

The ship sailed throughout the Mediterranean—Sicily, Malta, Tunisia—during its rebuild period. In November 1971, *Sofia* returned to Spain and tied up in Santa Pola, where a boat named *Sorcerer* was hauled out (*Sorcerer*'s swashbuckling captain, Nat Benjamin, once walked below decks through *Sofia*'s many smoky rooms, admiring the fine vessel, though he was not inclined to retain many memories afterward). When her rebuild was complete, *Sofia* headed to the Canary Islands off the western coast of Africa, and from there crossed the Atlantic to the West Indies, where Brad and his crew spent the better part of a year delivering cargo between Trinidad and St. Lucia, mainly construction material. They then moved on to Curaçao, then to Aruba, and sold a charter to some of their parents for $15 a day before heading through the Panama Canal and over to the Galápagos Islands, about 600 miles west of Ecuador. The Galápagos then, in the early 1970s, were untouched, unrestricted, and to Brad the most beautiful spot on earth.

Owners got on and off the boat during these travels, maybe fifty in all during the time Brad captained her; the incoming would pay $1,500 for a share in the boat and a say in her destination.

In the Galápagos, the boat sank in shallow waters after the garboards popped out of the keel. They got her afloat and limped to Costa Rica, where it took them nine months to repair her.

They sailed her next into the deep Pacific, to French Polynesia—the Marquesas Islands, Tahiti, the Cook Islands—and on to Tonga, an archipelago whose landscape ranges from volcanic mountains to lowlying coral formations. From there they sailed for Fiji, and then New Zealand, where they stopped to live and work for a year and a half.

Sofia set sail again, north toward Fiji, then west to New Hebrides, the Solomon Islands east of New Guinea, up through Micronesia, across the South China Sea through Indonesia, past Borneo, Vietnam, and Thailand, and up to Singapore, then through the Strait of Malacca, which separates Sumatra from the Malay Peninsula, then into the Indian Ocean, to Sri Lanka, the Maldives, the Seychelles, and Aldabra Island, then to Tanzania, Kenya, and Madagascar, then over to South Africa, around the Cape of Good Hope and up to St. Helena,

1,200 miles west of Angola in the South Atlantic. She then crossed to Brazil, headed up to Barbados, Bequia, and Martinique, bombed through the Virgin Islands, and then sailed from Tortola on up to North·Carolina and New York, at last anchoring in New Bedford.

And that was where Brad got off.

He'd flown to Sweden in 1970 (Vietnam at a boil, Nixon triumphant in the White House) intending to return two years later to finish college. It was now 1978 (the war in Vietnam over, Jimmy Carter president), and America was a different country altogether from the one he'd left. Linda, soon to be his wife, was about to bear her second child and their first together, Willow, and Brad by now knew one thing for certain: he wanted to remain a part of the sea. It was only a matter of months before he was in Denmark to buy another boat—a bigger one, with a steel hull, 100 feet on deck, 130 feet overall, 21 feet wide. She drew 9 feet unless she was filled with cargo, in which case her keel dipped 2½ more feet.

Edna had been built in 1916 to fish for herring in the North Sea, a drift netter, and she did so until 1935, when she was converted to a motor-sailing cargo vessel. She'd been laid up for five years when Brad bought her. He took her back down to Portugal, where he rerigged her once again as a sailing vessel, a ketch with a fixed bowsprit, a large mizzen, and topsails, and added a yard to carry square sails. He put on a new deck, built a new cabin, and rebuilt the hull.

In 1980 he set sail for the West Indies with his wife, their two children, five crew, and 35 tons of cobblestones, to begin a new life moving cargo under sail. From the West Indies, where Brad incorporated the business as Deep Water Ventures, they sailed for New England, loaded the boat with secondhand clothing and empty drums, and embarked for Africa by way of the Azores. It was late November by the time they left, bad-weather season in the North Atlantic, and they endured two weeks of storms. Winds blew more than 50 knots continuously for three days, and the seas got up to 40 feet—great big long seas. Brad ran with the storm, staying below the low pressure to avoid headwinds, moving at 6 knots·under bare poles. The rerigged *Edna* proved a worthy vessel in her first rough weather.

They sold the clothing and drums in Ghana, loaded the boat with hardwoods and handicrafts, and sailed for the West Indies. Once there, they traded locally and made New England–to–Bermuda runs carrying construction materials.

In 1982 they sailed *Edna* from Bermuda to Brazil but were unable to enter that country without a visa. Stymied, they discovered Suriname via an encyclopedia they carried aboard ship for the children, the younger of whom was now four years old. The country's range of cultures sounded intriguing; that it didn't require a visa made it inviting. Suriname would prove from the very beginning the source of their most lucrative cargo. On their second trip out, the wood was so plentiful that they filled not only the hold with it, but all the decks up to the rail as well. After a day at sea, however, at two in the morning, *Edna* began taking on water, and the crew had to hustle to take down the topsails and the flying jib and then start throwing valuable wood overboard.

Brad headed to Nova Scotia to overhaul his boat, loaded it with bricks, and returned to Suriname via the Virgin Islands for more runs, but these proved so profitable that they caused their own end: Brad had been wanting to return to the Pacific for years—he smiles gently now and says, "You don't need a reason to sail in the Pacific"—and the Suriname cargo gave him and Linda enough savings to trip over the Panama Canal into the big blue expanse with its miraculous speckling of islands and archipelagos.

They did more trading here, mainly ferrying wood from Costa Rica, Burma, and Indonesia to the West Coast of the United States. The crew could each earn about $10,000 per year from this work. The rigors of life at sea grew increasingly difficult as the children got older, though, so they and Linda would step off in Hawaii for stretches while Brad ran cargo.

The sailing itself was never without serious risks. Once, in the Pacific, Brad picked up reports of a typhoon south of their position. They were midway between the island of Ponape, in Micronesia, and Hawaii, and he was relieved to hear that the storm was headed west. But each day its position showed it moving north. Predictions continued to hold that it would move west, but it didn't. This was a scary situation: here he was with his wife and young children on a sailing ship hundreds of miles from land, and a typhoon with 170-knot winds was defying predictions and seeming to *follow* them, to be drawn to them. Each day he'd locate its position, and it kept moving *north*. Straight for them. This thing was far too big to outrun. He'd listen to the radio predicting a westerly course, and yet the position given, day after day for five maddening, stressful days, was north north north. Brad knew that if this typhoon decided to follow them and run them down, they'd die. It was far more dangerous than anything he'd ever seen. Eventually the radio reports amended their predictions: the typhoon was indeed heading north. It would, in fact, pass directly over *Edna*.

At that point you do what you can. You tie everything down, secure every spot on deck that can let water in, so that you'll float no matter what. The danger spots are the cabins and companionways, all the right angles imposed on the boat—they're the weakest points on the vessel, and you do what you can to secure them. What happens in extreme weather is that green and white water more or less swallows the boat, taking with it everything it can—the entire boat, if possible.

But as soon as the reports began predicting a northerly route, the typhoon changed course and headed west. Brad and his family didn't feel any weather to speak of.

In 1985, with the rest of the family safe in Hawaii, Brad loaded *Edna's* hold with hardwoods from various countries—ebony, teak, rosewood, nyatoh, merbau, and paduak—and set sail from Kuching, Borneo, heading for San Francisco, about 5,000 miles east. The crew was an international one and included Mark Witteveen, a Dutchman who cared for and repaired sails; Jack Risser, Brad's longtime first mate; Kevin Campbell, a South African and a veteran of his country's war with Angola; and a West German named Paul Kuhn. A few days and about 100 miles off the coast of Vietnam, in the middle of the South China Sea, at dusk, Brad was below in the engine room and suddenly stopped working. Something wasn't right; he could feel it. He went on deck. Others had noticed them already: two 35-foot launches, old fishing boats, that had appeared on the horizon on their port bow, not using their running lights. Then they noticed a third launch on their stern. One of the forward two began to make a move toward *Edna's* starboard bow, an attempt to surround her.

Thai pirates were known in these waters. They preyed mainly on boat people fleeing Cambodia and Vietnam and were notoriously violent. It was said that they'd rip out your gold-filled teeth, then shoot you; they raped women before cutting their throats. They never left anyone alive. (Posted on Ginny's door in the G&B shop is a card, in memoriam, with a picture of a young woman at the wheel of a boat, her name, Melanie Jones—no relation to Ginny—and the dates November 30, 1953–November ?, 1994. The card, from the woman's parents, explains that their daughter flew to Palau to crew and cook on a 57-foot Swan sloop, *Aphandra*. Melanie wrote in a postcard that they'd set out from Borneo on July 31. Nothing more was ever heard from her, the rest of the crew, or the boat.) In the middle of the South China Sea, Brad considered three fishing boats' moving to surround his boat to be an urgent and potentially life-threatening situation.

Edna was cruising at 5.6 knots, the wind about 20 or 30 degrees on her quarter, not a fast sailing position. Brad ordered first mate Jack Risser to alter their course, putting the wind on their beam for more speed, and to start the engine and rev it to 400 rpm. With their course altered and the engine cranked, the launch that was trying to move to their starboard side could not quite pass them, but all boats maintained steady pursuit. Soon they heard bullets in the air: they were being shot at with small firearms. Then explosions sounded from the boats, followed by splashes at their stern: the pirates were using cannon in an effort to disable the big sailing vessel. Brad ordered Kevin Campbell, the South African former soldier, to retaliate. Kevin ran below and returned with a .306 semiautomatic long-range rifle. One of the boats remained parallel with *Edna;* the others pursued from behind, now nearing 100 yards away.

Kevin lay down at the stern—a counter stern, a good position to defend from. "Fire at their waterlines," Brad ordered. He wanted to deter them without killing anyone, if possible. Kevin fired and fired, but after ten minutes the boats were still in their wake.

"Fire into their wheelhouses," Brad commanded.

Kevin quickly unloaded a full clip, thirty rounds, blowing out their wheelhouses.

At last the boats fell away. Brad kept *Edna* on that same course, her engine still cranked, as darkness fell. He kept the running lights off. When he felt confident that they were no longer being pursued, and the adrenaline had stopped throbbing in his arteries, he instructed Jack to resume their original course for San Francisco, two and a half months away.

The tension did not fully abate, however, because on this sail, unusually, they carried additional cargo: 12,000 pounds of marijuana. The sail proved uneventful otherwise, and they rendezvoused with a fishing boat, as arranged, 700 miles off the West Coast. They lowered Edna's sails and off-loaded the 6 tons of dried produce, raised the sails again, and headed for California, while the fishing boat motored toward Oregon. Brad's part of the dope smuggle was complete.

✳

Brad would captain Edna for another two years, but off and on, occasionally turning the boat over to another captain while he returned to Hawaii and his family. By 1987 the demands of the boat were running contrary to the demands of the family, and Brad decided to sell her. Edna was then the oldest steel-hulled ship carrying cargo under sail in the world. A licensed captain, one of the few women who held that title, bought her to charter and carry cargo. Brad was without a ship. He worked as a carpenter, he worked in boatshops, he worked as a fisherman. And he continued to import wood.

In 1987, the man who owned the fishing boat to which Brad had smuggled marijuana was caught. He gradually began to rat out all of his associates, anyone and everyone ever connected with him. Brad was in Singapore, a few years later, when he got word that he would be indicted on four counts of smuggling and distribution five years after the event. He first flew to the East Coast and contacted lawyers, then headed to San Francisco to turn himself in. He had good lawyers and a capable and lenient judge. By law, he could have been sent to jail for two years; instead he was fined $15,000 and sentenced to six months in the federal prison in Stafford, Arizona, to be followed by 2,000 hours of community service. The timing of his smuggling was significant: under new, stiffer drug laws, the crime Brad was convicted of, had it been committed after November 1987, would have carried a mandatory sentence of twenty years.

The year Brad went to jail, 1990, Edna was lost forever. In late February she was anchored off Atiu, one of the Cook Islands in French Polynesia, when a freak storm blew through the early-morning darkness, carrying her onto a reef at a severe dropoff near the lee shore of the island. She tore a 30-foot gash in her hull, and waves enraged by Cyclone Sina broke her to pieces against the rocks. By daybreak a barge was able to approach the sinking ship and carry her uninjured captain and crew to safety. *Edna*'s bow would eventually wash ashore, the last anyone ever saw of her.

✳

By the end of three weeks Brad has cut all the waterways for *Gazela*, as well as the planking and sternpost timbers for *Lynx*, the schooner at Rockport Marine, and the big timbers he's promised Mystic Seaport, which is finishing up *Amistad*. Cutting the boxed heart made that order possible and also resulted in 2,000 board feet of planking. He's shipped two containers, more than 15,000 board feet in all, to Fort Lauderdale. But the *Gazela* project has been put on hold, meaning that he won't be paid for this wood for up to a year.

Brad still has 30,000 more board feet to saw, but he needs to return home for a couple of weeks. Well aware of the haphazard nature of travel here, he confirms his flight the day before, but when he arrives at the airport, the place is deserted. The plane has left five hours early. No reason is given, and no other planes are scheduled for days. Brad must ride for nine straight hours in taxis and minivans and cross another river to reach the Georgetown Airport in Guyana to catch a flight. At the ticket counter he finds that the airline doesn't take Visa and he's $20 short of the airfare. The ticket agent loans him the money, and soon he's on his way to New York. He will arrive home fifty-six hours after leaving Paramaribo. In two weeks he'll be on his way back down again, thinking about wood and boats, forever engaged in, and by, the workmanship of risk in its various forms.

V

R oss Gannon spotted *Edna* unloading cargo at Tortola's commercial shipyard when Gannon & Benjamin Marine Railway was not quite three years old. Ross was charter captain of a boat called *Charlotte Ann,* and *Edna* was then engaged in its lucrative Suriname–Virgin Islands runs with its most efficient cargo, wood. Ross was impressed by the hefty vessel, and by Brad, and he was eager to try the unfamiliar timber Brad was off-loading. "Can you bring some of that wood to Vineyard Haven?" he asked. Brad thought for a moment, then nodded. He seemed so vague about it, Ross didn't know if he'd see him again, but the following July, 1983, the 100-foot steel-hulled ketch sailed into Vineyard Haven harbor to off-load G&B's first purchase of Surinamese timbers.

G&B was building at the rate of almost one boat a year then (the catboat would be launched that year, the third Canvasback the next), and because the proprietors of the operation hadn't developed a fleet to repair and maintain and were largely unknown outside the wooden-boat-yard community (on the island, G&B was "that hippie boat-yard"), they were at liberty to sail for a good chunk of the winter in a chartering capacity.

In the summer many of the G&B crew lived in the harbor. Gretchen had a boat, Ross had *Urchin* and then *Undina,* Jim and Ginny Lobdell lived on their *Malabar,* and most every night one family or another would invite the others for dinner. Ross and Gretchen rowed to work. Folks sipped coffee on the dock before the day began, waking up there. Gretchen smiles at the horizon when she thinks of those days, when G&B was "just a shack on the beach." She worked part-time as a hostess at the Black Dog her first summer here, 1980, and so was able to shower there. She was in her midtwenties then, and she loved this spot—the Black Dog, the boatyard, and the P.O. across the street were, she says, "my little world." Nat was more or less in the process of settling down, with the aid of a nightly six o'clock phone call from Pam to ensure that he was headed home and not to the Ritz, the shabby, comfortable bar where everyone went to drink and draw boats on napkins and tabletops. And the business was if not huge and lucrative, then at least stable. Neither Nat nor Ross could say if they'd have work from one year to the next, but somehow they always did. And Ross was finding that he really *enjoyed* building boats—they were endlessly fascinating to him. Their beauty gave him pleasure, and at the same time, they satisfied his need to flip, rotate, or transport heavy objects.

In 1986 the boatyard received its first attention in *WoodenBoat*, after an art professor and deepwater sailor wrote about his 1929 schooner's repair at G&B, "chosen for [its] reputation for putting new life into old but potentially good hulls."

The writer, Peter Phillipps, went on to describe Nat and Ross in those early days, 1984:

> Once the boat had been chocked on the railway, Ross Gannon slipped below. He smiled politely, and with hammer in hand immediately began in earnest, standing on the bilge stringer and walking the 30-foot length of the cabin. He stopped at each cracked frame, laid his head against the planking, and tapped his hammer at specific spots. He did not seem to hear my nervous questions. "Too bad they kerfed these frames and did not back-fasten them," he said softly and with compassion. . . .
> I suggested meekly that this might be a good opportunity to replace, maybe, all of the buttblocks with oak. His eyes moved to a vulnerable mahogany buttblock next to his hammer, and with the claw he struck, split, and removed the upper edge in one blow. Then with six sharp shots he drove the fastenings outward. No discussion necessary.
> . . . Initially it seemed brutal, but Ross's technique was actually effective and quick. The manner with which he and Nat attack a situation is always expedient, once a course of action has been determined.

Phillipps also described G&B generally: "It's a family yard, one where local boat owners often stop by and, whether they are dressed for it or not, invariably find themselves unloading a truck of lumber or picking up a sander and slowly fairing someone else's hull." (G&B still takes frequent advantage of the boatyard's Pied Piper effect, amassing hours of free labor from people who can't quite remember why they stopped by in the first place. It's how the yard earned its nickname, "Grab 'em & Bend 'em.")

During those same years, Nat and Ross bought a boat of their own, *Zorra,* a 72-foot Italian-built yawl that in a way symbolized the propitious fortunes of G&B in the 1980s. They spotted an item in the *Boston Globe* noting the auction of a big and damaged wooden boat. Neither man had heard of her, but an old-timer who happened to be at the shop when they were discussing her had, and he told them she was a fine vessel.

They went to have a look. The interior was black from an electrical fire, but when they rubbed away the soot, they found beautiful mahogany, beautiful brass fixtures. She had been well built, they saw. The insurance company was selling her as totaled, but Nat and Ross figured they could repair her within a year. Ross put in a bid of $38,000 at the sealed-bid auction, and they owned a boat. (When they eventually sold her, in the mid-1990s, she went for more than a quarter of a million.)

They had the sexy white-hulled rocket sailing in a year, and she bombed up and down the New England–Bermuda–Caribbean route for the next half-dozen years. Her finest year, though, according to Ross, was her first with them, before an engine was put in. That was his happiest sailing, because it required all kinds of problem solving and expertise to maneuver that big boat in and out of crowded harbors. *Zorra* determined the shape of those days for G&B because her repair, her maintenance, and the work of chartering her took up a good chunk of each year. And because getting her ready for the charter season always took longer than planned, *Zorra* often left Martha's Vineyard in early winter when the weather is unpredictable. A few years after the engine went in, Ross took the boat south in early January with his nephew Antonio, a friend named Malcolm Boyd, and a half-dozen others who came along for the sail. Ross and Antonio had spent two months in Fairhaven fitting her out for the season (she was too big to haul out on the Vineyard). Ross came aboard at the last minute, he and Nat having had second thoughts about the skipper they'd hired. The second day out, a nor'easter struck. The wind built all day, and by evening all sails were down. The seas were short and steep. After sunset it really began to blow, and looking to weather was difficult. They ran downwind, surfing the waves. Antonio was on watch with Malcolm. Waves began to crash over the stern. There was a life raft well lashed to the cockpit sole; a wave ripped it clean off, and they never saw it again.

"It was very well lashed," Ross remembers, "and we were repeatedly taking waves over the stern. Heavy winds were blowing out of the north. And we were going like *mad* under bare poles—we were going about ten knots with no sails. It was absolutely pitch-black night. And then it got really nasty when a wave knocked out the compass light."

Antonio had to work hard to steer this big boat. Little was visible now other than white water all around them. Antonio felt the boat suddenly rise, and he turned to see behind him, and ten feet over him, a wall of water just as it crashed over the boat. It sent Malcolm flying across the cockpit sole and Antonio into the binnacle and pedestal steerer, made of cast aluminum and bolted to the deck. It broke off at its base, and Antonio lay in a tangle of lines, one of which was his own lifeline. "We lost steering!" he shouted to Malcolm, who relayed the news to Ross below. Spreader lights flickered on to illuminate the damage.

Ross came up on deck, calmly, and got the binnacle back up, securing it with a triangulation of lines attached to winches. The wheel was still connected to the rudder, but steering was impossible. They could do nothing but wait for this weather to pass—40 hours ahull in 70-knot winds. The hired skipper panicked and, unbeknownst to anyone, sent out a Mayday—something you do only if you are sinking. An angry Ross informed the Coast Guard that *Zorra* was without steering but did not need assistance. They were blown 150 miles sideways. When the weather abated, a Coast Guard cutter came to check on them. By this time they were securing the binnacle so that the compass would work and they could steer using a fenderboard.

They staggered into Bermuda a few days later. It took Ross and his crew three weeks to repair the damage so Malcolm could skipper *Zorra* south for the charter season.

✳

These early days were not without adventure and drama, but more frequently they were composed of productive work, good boats, and good sailing on Martha's Vineyard. Three Canvasbacks built, a catboat, *Swallows and Amazons Forever, Liberty, Lana & Harley*—and the yard's reputation for expert repair and restoration and for building beautiful wooden boats grew.

By the fall of 1989 they were at work on a number of boats. They'd just finished a 10-foot sailing dinghy that needed only one more coat of paint before the owners took her to the Caribbean. They had a salvage job on the rails—*Java,* a 57-foot yawl that had sunk and then been given to them; they'd thought it would be fun to turn her into a boat again, and she was on her way to being a fine schooner. Beneath the overhang was an Ohlsen 35, and also hauled was a small cutter named *Corineus,* originally built in 1927. But the most dramatic of the works in progress was a 33-foot ketch being built in the outer shop for a man named Al Kent, over on Cape Cod. Nat's new ketch was coming along beautifully. She was planked and caulked, with deck beams and bulkheads and cabin in place, cabin sole, cockpit sole, interior, and beside her a big pile of teak ready to be sawn and turned into a fine laid deck.

The weather was turning cool in mid-October, but only just, since summers linger here, and this Monday had been a good beginning to the workweek. Ross was living with Suzy then; she was four months pregnant with Lyle, and Ross was determined to make this relationship work. He'd eaten and was so tired he fell asleep on the couch before crawling upstairs to bed. Nat slept, too, having put in a good day on the new boat, but Pam, unusually for her, couldn't get to sleep. She took a walk down Grove Street to the beach at around midnight and saw there an ominous orange glow. She hurried back home. Vineyard Haven fire alarms had begun sounding. Shortly after she reached the house, an explosion sounded in the distance. The phone rang almost immediately after the boom.

The phone rang next at Ross's. Kristi Kinsman—who lived with Gary Maynard above the Black Dog, near the boatyard—breathlessly and quickly explained that the shop was on fire.

Nat was already out the door and speeding down Main Street. Gretchen, who at that time lived on Hatch, just down the street, pulled out just as Nat passed and followed him. As soon as Ross was out of the house in West Tisbury, he saw the orange glow in the direction of Vineyard Haven and understood immediately the magnitude of the fire: catastrophic. While he and Suzy barreled down State Road, the flames reached a hundred feet into the sky, the outer shop already gone. They made it in time to see the skeleton of the new boat, black against the orange flame. The planking had already burned away. Then the frames blazed brightly, burned out, and fell to the ground like a stationary fireworks display. The bow and the stern tipped and fell, too. Nat and Pam, Gretchen, Ross and Suzy, Kristi and Gary—they all just stared, lit by the glow, surrounded by the noise of flame and crackling timbers.

Someone noticed that the tenders pulled up on the beach were starting to ignite, and this broke through the shock. They all now ran to the beach to haul the tenders into the water, to save anything in danger that could be saved. The 1927 cutter *Corineus,* on the rail, seemed safely away from the fire, but suddenly flames simply ignited on her hull from the heat. There was no electricity to run the winch and let her roll away into the water, and everything was powerfully hot, but Nat and Ross were able to get to the winch, and several others heaved on the cable to create enough slack to remove the pawl, and that freed the winch, allowing the boat to slide safely into the water.

Ross's boat *Undina* was tied stern to the dock, and he let the line out to put another boat-length between her and the fire. A southwest wind pushed a shower of ash and soot into the harbor. Nat, strangely, didn't seem much different from his usual self. Not that he wasn't disoriented, too—nothing was computing; this didn't make sense to him, wasn't real—but there was a calmness there that seemed a little odd. He didn't appear all that concerned about the loss of the new boat he'd designed and had been building for half a year, now turned to ash; or the loss of the shop; or the loss of the business. What truly upset him, what would always remain the greatest loss in his eyes, what wrenched his heart most, he said, was the sight of all those *trees* burning. Gretchen was in shock and disoriented. She'd been in the process of moving, and most everything she owned was in her loft, including cherished photo albums. Ross felt devastated and blank. Eventually the fire was brought under control, and the man in charge told the group that the fire team would keep an eye on things through the night—he suggested that they go home and get some sleep.

No one slept. Ross lay on his back in bed and stared at nothing, thinking not one coherent thought, stunned by fire. Gretchen could think only about all the things that were gone, running through an inventory in her head. She waited for the sun to come up and then returned to the field of spent, wet embers and ash, a 5,000-pound puddle of lead cooling where Al Kent's boat had been. It was a rainy, cold day. Nat and Pam came not long after, and then Ross. That was when the pain set in. Sifting through the ashes in the rain, registering the magnitude of the loss piece by piece, their entire business gone. What would happen?

Nat felt in his gut that they'd build a better shop and a better boat-yard, but he didn't know if that was really possible. Ross didn't know, either. They were not very well insured. Any money they got would go to cover the boat owners' losses. Gretchen thought maybe it was time for her to leave the island and start something else somewhere new. All around them they saw loss: their old hand tools—special pieces from the early part of the century, designed for specific boat-building tasks—as well as the great old nautical hardware scavenged here and there, pieces you just couldn't find anymore, and of course the heavy machinery.

Then the people began to turn up—a stream of ambulance chasers in trench coats, people looking to make a buck off someone else's disaster, insurance investigators trying to nose out if maybe the owner hadn't been able to make payments on that expensive boat that had just burned up.

But the local fire inspector arrived and quickly pronounced that arson did not seem to have been the cause, and told them they could begin to clean this mess up.

Then islanders started to show up, too, locals asking to help—ten of them, then twenty, then fifty, all sifting through the ash and the charred scraps of wood and melt for anything salvageable. Ralph Packer had a Dumpster delivered for the garbage. The Black Dog sent over boxes and boxes of doughnuts and an urn of coffee for the volunteers. Gretchen, Ross, and Nat were all grateful, their gazes cast downward.

But the people didn't *stop* coming the next day. Messages poured in from all over, the post office box and answering machine filling up and filling up. Money came, too—not fives and tens but checks for $50 and $100 and $500. A friends-of group was quickly formed to organize the work: setting up a bank account for donations and planning benefits to raise money, one at the Black Dog and one at which some of the Taylors—James, Kate, and Huey—would perform. Ham Fish, Nat's cousin, wrote fund-raising letters and moved the message out into political circles, where there were many prominent lovers of wooden boats.

Bob Douglas found room in his next-door offices for Ginny to set up a temporary office, plug in a phone, and begin to reorganize the business. Gretchen was given space across from the Steamship Authority that she could use as a temporary loft. Erford Burt, the octogenarian designer of the Vineyard Haven 15s, came by to donate some of his tools, ones he knew Nat and Ross would appreciate—a Nova Scotia adze with a lip on it, a broadax with an offset handle, special hand drills. Bigger tools came over from Maciel Marine. Cheap ship's saws and planers were offered. The boatyard at Mystic Seaport in Connecticut took up a collection of spare tools from all their boatwrights and shipped them to Vineyard Haven. Letters and condolences and checks, hundreds of them, continued to flood in.

The message from the community was clear. Nat and Ross couldn't give up even if they wanted to; they *had* to get back up and running. They didn't have a choice.

Neither Nat nor Ross had had any idea that so many people were watching them. They had both thought they'd been quietly building all these old-fashioned boats and nobody really cared, or even *noticed*. And that was *OK*. This astonishing outpouring from strangers was almost too much to bear. The gratitude was fraught with guilt; it's not easy to accept such a torrent. When Ross appeared meekly (he didn't want to go at all) at the Black Dog benefit more than a week after the fire and saw that his mother had flown up from Florida to be there as a surprise, he couldn't maintain control any longer, and he broke down and wept in front of all those people.

✳

Within a week Ross had a plan in his head for a new shop, wood had been ordered from Jim Aaron, and an island boatshop raising was scheduled. Nat and Ross heard that people were going to show up to help; they expected twenty-five. When Nat arrived at eight, the yard was, he says, "an anthill." More than a hundred volunteers had appeared. Seven of the island's contractors showed up to marshal the forces, both their own crews and the scores of others who just wanted to help.

"It was *un-believable,*" Ross says.

The frame was up before the morning was out—a brilliant, clear November day. "The speed at which this was going on was unbelievable," Ross explains. "It was sort of like feeding the lions. You'd throw down a board, and in a flurry of activity, sixteen people would nail down a sixteen-foot board in about three seconds. The hammering was deafening. They were still nailing down the boards on the first floor when boards were going down on the second floor."

A big pot of chili was brought to the beach, and pizzas were sent over. A bonfire was lit and mulled cider served. The work went on through that Saturday—Nat and Ross could see it that day, their new shop, or *the island's* new shop, really; it wasn't theirs anymore—and continued into Sunday. On Sunday Lynn Bouck came by and spoke for himself and his crew: "We'd like to shingle the roof tomorrow. Do you have the shingles?" And that was how the roof got finished. After one long weekend, Gannon and Benjamin were back, stronger than before. And with an enhanced sense of the community in which they found themselves. *They'd had no idea* of the magnitude. Because of this they felt a greater responsibility. When you understand what you can lose, you get serious. They had to run this less like a shack on the beach and more like a proper boatyard. Or, try to, anyhow. And they had a good structure now, one that wouldn't shudder when the nor'easters shrieked across the harbor. Now, a decade later, the point that Nat and Ross both stress when recalling the fire is how powerful and important it was for them to know how much a part of Vineyard Haven, of Martha's Vineyard, they were. Ross says of the fire that far from being unfortunate or devastating, it was in fact one of the best things that had ever happened in his life. A gift.

It wasn't a gift to Al Kent, of course, whose brand-new, *almost finished* boat was gone, gone, gone. He was about to retire and had already been sailing his new boat in his mind, dreaming of the coming spring. And now he had no boat. Nat and Ross, for all the speed with which they returned to business in their spanking new shop, had to tell Al they couldn't begin rebuilding his boat right away, certainly not in time for spring. They would understand if he went looking elsewhere.

And he did. But after about a month he, too, came back, and said, "I want you guys to build it." Which they would do, bigger and better than the previous version. *Encore,* Al Kent's 37-foot gaff ketch, was launched the following year off the G&B beach.

The gifts of the fire proved substantial, solidifying G&B as a business within the Vineyard community and reminding the wooden boat world through this remarkable story (fully chronicled in *WoodenBoat,* itself once galvanized by fire) who Ross Gannon and Nat Benjamin were, and what people felt about them.

Two new boats followed the fire—*Encore* and then the yawl *Candle in the Wind,* which Joel White would critique so lovingly in *WoodenBoat* after her launch in 1991—but after these a scary dry spell hit, four years of repair work and not a single new construction, though the yard was buoyed by the reconstruction of that black-hulled Percheron of a schooner, the 86,000-pound *When and If.* Then a Columbia University linguistics professor named William Diver commissioned a boat called *Tern,* a modified Rozinante, and with that launch in 1995, the new designs and constructions resumed in unbroken and increasing succession. And that in addition to more maintenance work on all the boats G&B had been building and more repair work on boats that came to the yard as its reputation grew. The work in fact became daunting. There was almost too much of it. You couldn't work with a drawl in your movements anymore. There was no more coffee on the dock in the morning while you stretched the sleep out of your body. You had to barrel through the work behind Ross's interference. And all of it, this increased work and new construction, rode the resurgence in wooden boat construction of the 1990s.

In 1998 Nat began work on the greatest of his designs, greatest in every measurable way—weight, size, complexity of construction, aesthetics of design. *Rebecca* would be the biggest plank-on-frame vessel to be built on the Vineyard since the mid-1800s, a 60-foot sea creature with a mighty bow and a swift, graceful stern, a marconi main and gaff foresail schooner, a deepwater vessel built to cruise the Pacific and the Indian Oceans alike, built powerfully to turn the notorious corners of Good Hope or Cape Horn, built to be comfortable and safe in the late-autumn blows of the North Atlantic as she headed south for the winter or across to the Azores. Nat, now in his early fifties, beloved at home and in the community, heading a solid business, completely at peace with himself and his work, was arguably working at the height of his powers at the exact moment the commission for *Rebecca* appeared. She might be a vehicle for all his ideas, experience, and knowledge; he might put his entire being into the service of this boat. He might give all that he knew, like any artist, and all that he'd learned in his half century of life, to this boat, a realization in wood of those helical symbioses of art and science, safety and beauty, and durability over decades. *Pieces of wood held together by bronze that stay together in dynamic conditions at sea.* You couldn't kill it. You couldn't kill this thing if you tried, not G&B on a beach in Vineyard Haven, not *Rebecca*, not the wooden boat species. The wooden boat, big or small, after all, embodied and gave extended meaning to the natural world, trees harvested one by one, sawn, planed, bent, and faired with small tools into an exquisite shape that could be both home and vessel, that you could sail away on. *Rebecca*. A big vessel built by hand—art and science, uncertain, imperfect, and bound for the beautiful, deadly wilderness of the sea.

Five

Rebecca, Elisa,
and *Jane Dore*

I

On February 8, 1999, fifteen months after five big trucks rumbled down Beach Road, each carrying 20 tons of Suriname hardwood, work had stopped on *Rebecca*. Since that day, she'd remained a sealed hull, two deckhouses fashioned, big beams of oak framing the deck, silverballi already stickered beside the shed and ready to be ripped into deck planks. The interior had been about to go in, berths, galley, salon, the boomkin and bowsprit and rudder about to be fashioned: eight more months of work left and then a September launch, that was what everyone thought—maybe sooner, given the head of steam the whole crew had going, and the excitement of the work. Half a year. That soon. After all she'd been through, *Rebecca* was so close to being a boat that you could almost see her slipping into Vineyard Haven Harbor, joining the likes of *When and If* and Mystic's *Brilliant,* two extraordinary schooners built in the 1930s that shared *Rebecca*'s art and scope, her rugged construction, her elegant lines. But that fall launch would not happen.

Dan Adams would confide to me that he'd seen it coming months earlier and, unbeknownst to Gannon & Benjamin, had been struggling to get construction loans to avoid what ended up happening on February 8 at Mugwump. His then girlfriend and soon-to-be fiancée, Priscilla Wrenn, a local real estate agent who sold him his Chilmark barn, shook her head pleadingly and explained, "The part of Dan that people don't understand is the idealistic part of him."

Perhaps his idealism, then, had gotten in the way, or perhaps he'd been caught in one too many financial maneuvers. There were, Dan

told me, "a lot of circumstances." The situation should have come as
no surprise to anyone, especially after that time Ginny went to send
money to Brad down in South America, only to find that Dan had
cleaned out the bank account. This was not an isolated event. Rumor
and fact began to form a picture of Dan's financial untrustworthiness
generally. It had been reported more than a year earlier in the *Cape Cod
Times,* for instance, that he still owed $24,000 for motel bills and a
house he rented during the filming of *The Mouse*. Another of the "circumstances" involved a former girlfriend who put a lien on "everything Dan owns," Nat said, "and a few things he doesn't!" One of
those things, regrettably, was *Rebecca*.

Liens notwithstanding, rumor had it that Dan would soon have a
loan to complete the boat. It made sense—the boat had to be worth
more than half a million dollars, enough for collateral. Throughout
late winter, spring, and summer, the words floated through the shop,
aloft on the buzz of the ship's saw and the rattle of the planer. "A loan
is supposed to be coming through at the end of the week," it was said,
or, "Not this week but next—no, this time it's true, Nat talked to the
appraiser." February. March. April. "No, really, this time it looks like
it'll go through." May. June. July. August. September—the month she
was to have been launched. September comes, race month on the
Vineyard, the month the island natives take off after the heavy summer
season, the month that signals the coming of fall and another year,
back to work at the yard. Mugwump remains deserted.

Or mainly deserted. David Stimson does occasional work there.
It's more his home than the G&B yard, his employment at which, due
to rumors of the impending loans, he has had reason to believe may be
temporary. After the *Rebecca* shutdown he built a skiff at Mugwump
for a camp in Maine that he's often done work for. He keeps his tools
there, too. Of course, a steady stream of curious observers still trundle
through the shed, having heard about the schooner and wanting to
have a look. It's a remarkable sight even in still-life form—not often do
you see a big boat like this in three-quarters-finished construction,
every plank, every double-sawn frame, every floor timber visible.

One fine spring day, I joined Nat and David to head over to Mugwump to help David load the new skiff into the back of his truck. Fred
Hecklinger, a marine surveyor from Annapolis, Maryland, and emis-

sary from yacht designer Melbourne Smith, had dropped by the yard and spread out some drawings on the table saw for Ross and Nat. A gentleman in California had commissioned the design and half model for a big boat, a 90-foot schooner, 70 feet on the waterline, to be used as a yacht Coast Guard—certified to carry paying passengers. Fred wanted to know if Nat and Ross had any interest in building it. He walked out with Nat and David Stimson and me as we headed to Mugwump, and he and Nat continued to talk about the 90-footer. Ross definitely wasn't interested—G&B didn't have the time (three years), the space, or the manpower to devote to such a build, he said—but Nat wanted to consider it; the only obstacle, he felt, was that they had no place to build it, certainly not with *Rebecca* stalled where she was. Fred knew that Nat knew that yacht building was often a thing of people's dreams, and the drawings he carried were an extraordinary kind of dream, so he told Nat there was no balance due for the design and half model. And of the Californian's financial position with regard to the entire project, he said, "He shows every sign of being able to pay for it."

Nat turned to David and said, "Ah, we've heard that before, haven't we, David!" He laughed hard and heartily. Fred said good-bye and we carried on to Mugwump to finish our errand. When the skiff was in the truck, Nat pulled back the shed's door flap, which remained lowered but unsecured—it was not yet padlocked at that point—saying, "I'm just going to have a look at the temple." Then he added, "The temple of idleness."

He walked around *Rebecca,* craning his neck to gaze at the hull, and David asked, "Have you heard from him?"

"No, not a peep."

The most recent rumor had it that a bridge loan was on its way to provide money till the *real* loan came through. As Nat and David moved toward the stern, both started slightly at the sight of a silent Dan Adams, leaning on an overturned tender on sawhorses beneath *Rebecca's* stern, a Black Dog cup of coffee in his hand, gazing up at his boat. Greetings were exchanged, followed by the inevitable question from Nat as to the status of the loans. Dan explained only that a bridge loan would be very expensive—"I was hoping not to do that," he said. Then he said something that clearly took Nat aback. Dan noted the usual barrage of spring work at G&B and asked, "Are you all caught up?"

Nat's head jerked slightly. Then he said, "Yeah . . . ," tentatively, and noted that "catching up" was not really an issue—this work happened every spring.

But Dan said it again: "If you're all caught up, maybe we can get a few people over here next week." This seemed spooky to me—the guy had made me uneasy from the first time I saw him, and now he was trying to manipulate Nat's perception of the situation in a peculiar and meaningless way.

"We'll get back to work as soon as we're *paid*," Nat replied, annoyed. "We're *dying* to get back to work. David here is dying to get back to work, and as soon as we get paid, we will."

David excused himself uneasily to put padding around the dory in the truck. But the conversation quickly turned neutral, first to the renovated barn (Dan had put it on the market), then to Tori, his erstwhile girlfriend, who he claimed was spreading scurrilous lies about him, then to the new movie that he hoped to begin filming in the summer.

"Have you found a female lead?" Nat asked.

Dan named her and said she was an up-and-comer. Dan noted other possible participants, including his friend Armand Assante, and Nat said, "Why don't you get Armand to give you a loan?"

Dan shook his head and said, "He'd want something for it. He'd want a month in the Caribbean." Again, the disregard for logic seemed eerie to me.

"Tell him he can have it!" Nat said.

"How's Jonathan's boat coming?" Dan asked, changing the subject.

"It's coming."

"I just heard from him; he's going to be here."

"He's supposed to be here the seventh."

"Right," Dan said, then asked, "Is he still planning to do the interior himself?"

"No."

"No?"

"No, he wants a lot more done to it. Hot and cold running water."

"Really."

"Refrigeration."

"Really."

The list of additions went on, and Nat admitted, "It's going to practically double the cost of the boat."

"Does Jonathan know this?"

Nat said he and Ross had gone over it with Jonathan, but he hadn't yet seen an actual bill.

After a few more gentle questions about the loans, Nat concluded on a friendly note: *"Well.* We'll have a launching a year from now, as we originally planned."

Dan nodded.

Nat cleared his throat and said, *"Well,"* which meant, Back to work.

No one saw or heard much more of Dan Adams all spring and summer.

✴

September is lazy and ragged at the yard as everyone recuperates from the difficult summer and enjoys the best weather and sailing of the year. By the end of September the nights have begun to cool, and during the days, the boatyard readjusts to its fall-winter routine. The yard is restoring an old catboat for a historical society, iron fastenings and all. G&B has been given a power launch that needs a rebuild—the boatwrights will do that work and use it for their launch service. And Ross has had the unexpected pleasure of picking up a boat called *So Long* and hauling her back to the shop on his trailer. *So Long,* one of the few remaining Vineyard Haven 15s, was the first boat he owned upon arriving on the island in the late 1960s. He'll restore the sixty-year-old boat to bright new sailing condition for her new owner.

One difference this year is the institution of weekly meetings to ensure that everyone works as efficiently as possible. On Monday, September 27, Ross holds court inside the shop shortly after eight; in a circle around him are Ted, Kerry, David, a new man named Mark who began in the summer, and Bridger, a cheerful teenaged apprentice who lives in a tiny plywood shed in the brush of Bargain Acres. Ross scans their faces, takes a breath, and says, "Well. It's going to be a slow day today. Should I just go around and assign jobs?"

Ross is depressed because Kirsten has been asked to help deliver a boat from Brazil to Tierra del Fuego, the archipelago at the southern tip of South America, with a brief stop in the Falkland Islands; he's wishing she'd settle down. His whole demeanor is downcast: he's never been so undone by a woman, and it's driving him crazy. But there is work to be done, and he can lose himself in that. David Stimson says he was planning to organize tools at Mugwump, but Ross tells him to work on *Elisa*'s interior, instead. Jim, Ross says, should patch the deck of David Bramhall's boat, which they'll bring to the dock. Ted, Mark, and Bridger will work on the power launch; he instructs them to drag some oak out of the shed, start cutting frames, and fire up the steamer. Ross tells Kerry, the ex–cranberry farmer, who's sipping coffee from the cup of a silver thermos, to paint the hatches for the powerboat and maybe get to work on the dash when he's finished. As for Ross himself, he'll work on *Elisa*'s rudder.

He concludes the meeting with news, saying, "The situation with *Rebecca* . . ."; then he pauses before going into the story.

On Thursday, he explains, Dan Adams called the shop and requested that Nat and Ross get their paperwork to their lawyer because a loan was going through. Ross figured it was more of the same, but they're obligated to comply, and so they alerted their lawyer and got him the requested paperwork by Friday morning.

Ross now pauses again. "And it happened," he says. "Dan got a loan on Friday."

David Stimson's mouth falls slack, and he says, "Unbelievable."

Kerry says, "Wow."

"So," Ross says, "if he spends the money on *Rebecca,* it *will* go forward sometime this fall or winter."

Ross notes that there's still considerable paperwork to slog through, and a lot of ass covering to be done, something they didn't do enough of in the early stages of the project. He reminds everyone that Dan still owes the shop money, and adds that Dan claims it's the other way around, but he says the differences don't seem insurmountable.

Ginny has stepped catlike from her office to listen in and, referring to the new money infusion, asks, "Does that cover the vendors?"

Ross responds, "No, just us." Ginny's mouth tightens—she's the one who deals with unpaid suppliers, many of whom are also friends, all of whom the boatyard used to enjoy good relationships with. Ross promises, "We won't work until all vendors are paid."

✳

And so in late fall, after three quarters of a year of no work, the boatwrights gradually file into the Mugwump shed to pick up their tools and carry on where they left off, largely the same crew. Much work remains to be done, notably the boat's interior. Nat has designed the space below decks to accommodate eight comfortably in a conventional layout. Two companionways are to connect the entire area below, which will include, running aft to fore, the navigation station, the owner's cabin (double berth, bureau, private head and shower), a galley, the main saloon, a second head, and two berths in the fo'c's'le. Construction of the deck will follow the interior. Laying the 1¾-inch strips of silverballi and teak that will bend around the length of the waterways in what's called a sprung-laid deck is among the first orders of business. It is possible to lay strips of decking straight fore and aft, but when bent to the curve of the sheer and fastened at either end into nibs notched out of a kingplank, the deck becomes a thing of beauty. And according to Nat, this has proved over time to be the best kind of deck construction there is, both watertight and easy to repair. "Old-fashioned," Nat calls it. "Traditional and stronger."

The work of laying a deck is gratifying in the same way planking is gratifying. It's the last big visual change in the boat until the masts are stepped. Decking encloses the boat, creates two levels, above and below, and gives you a·way to walk its length without having to hop from deck beam to deck beam. The bottom edges of each strip of wood are chamfered, so that the overhead is grooved. When the wood has been fastened with bronze screws beneath silverballi bungs, the deck is sanded and then, like a hull, caulked with two layers of cotton and sealed with rubber compound.

As work on *Rebecca* proceeds, however, Dan Adams's financial affairs began to contribute the now-expected elements of drama. The crew works through the winter, through the big New Year, and into 2000. But by mid-February Dan's management of the project has proved so inefficient and, according to the lender, "dreamlike" that the lender steps in. He is Pete Normandin, a Massachusetts businessman and manufacturer whose company, Ranor, built more nuclear-waste containers than any other in the world. These "nuclear trash cans," as he calls them, are, as Vincent Scully pointed out, one of those few items (like *Rebecca*) that we build to last more than a generation ("I don't want to get into a pissing contest," Normandin says of his cans, "but they'll outlast any boat G&B builds by far"). Normandin's retirement sideline is lending money to people for interesting, unconventional projects. He was fascinated by Dan's pitch for *Rebecca* and happy to loan him the money he needed, but eventually his debtor's continuing inability to make his payments forces him to take charge of the schooner. Nat, Ross, and Ginny all like and trust Pete, a straight shooter from working-class roots who's been coming to the island twice a month to oversee the boat's progress, and so they're delighted by this development. Dan apparently accepts Pete's move even though it's the first step in his possibly losing the boat, but he'll attempt to raise the money necessary to win back control. Until then, though, Pete is running the show, and Pete guarantees that all the crew will continue to receive paychecks. Furthermore, it would not disappoint Pete in the least if he wound up the owner of this vessel. He isn't a sailor, but he's always loved and owned boats, and he's become captivated by Nat's work and wooden boats in general, and mesmerized by *Rebecca* in particular.

As Dan's "house of cards," as Ross calls it, begins its slow-motion collapse (the brick barn, which he's been trying to sell for more than $1.5 million, will also be foreclosed on), Nat quietly starts to put the word out that *Rebecca* may soon go up for public auction. Several wealthy businessmen travel to have a look at the boat, people such as Henry DuPont, of the DuPont family, and Robert Soros, son of the hedge-fund billionaire.

So in April 2000 construction of *Rebecca* bombs along under Normandin's smooth financial management, ahead of schedule for a summer launch, and under budget. Nat begins working on the patterns from which he will fashion an 8-foot-tall rudder, as well as patterns sculpted out of pine to be cast in bronze for the rudder's hardware. He finds a blacksmith on the island who will be able to hammer out a good gammon iron for the stem. A bowsprit of solid yellow pine has been made, more than 14 feet long and 9 inches thick at its heaviest sections. Casson is installing the boomkin, also made of yellow pine, comprising two long timbers that extend off the stern nearly as far as the bowsprit will extend off the bow. David Stimson and Todd McGee both work on the caprail, the piece of wood that will ultimately define the sheer. Brad has begun to wire the boat; a daunting tangle currently hangs out of the circuit board. Buckaroo Robert, Kirsten (returned, to Ross's delight, from Tierra del Fuego, and now pregnant with—surprise!—twins), and G&B friend Simmy Holzer do finish work below, sanding and varnishing woodwork. There's still much to be done here, including completing the heads and building shelves and hatches. The spars and masts are to be built by Antonio, Ross's nephew, who drew the construction plan and ran calculations. And Billy Mabie—"No cash, no splash!"—has officially been commissioned to do the rigging.

At the end of March, Dan faxed a cease-and-desist order to Nat, demanding that he halt work on the schooner. Nat dismissed it as just more of Dan's shenanigans, but two weeks later Dan declares Chapter 11 bankruptcy on behalf of Mugwump Charters, Inc. Chapter 11 allows the filer to retain some money and possessions while paying some creditors gradually. This will buy Dan time in which to reorganize his finances and attempt to reclaim the boat, and it forces Pete Normandin to instruct Nat to stop work. Dan soon changes his bankruptcy plan to Chapter 7, a more sweeping form that requires him to liquidate all his holdings to pay his creditors. All of this has to go through bankruptcy court in Boston, and in the meantime work on *Rebecca* is halted once again, mired in Dan Adams's grasping efforts to keep his boat.

By Nat's estimate, at the time of the second shutdown, in mid-April, *Rebecca* is ten weeks shy of a launch date. Everyone seems fed up by this point. The delusion of Dan Adams has become too much even for the generous and philosophical Nat to bear any longer. Shortly before the shutdown, but after the cease-and-desist order, Dan came to the boatyard to have a look at the progress and to see what more needed to be done. Nat told Dan he was no longer welcome here and asked him to leave, pointing out that it was G&B property. When Dan refused, Nat called the Tisbury police. After the shutdown he asks the police to keep an eye on Mugwump and the boat at night. He's seen boats burn before, and he doesn't want to take any chances on anything happening to *Rebecca*. Soon, on an island whose natives rarely lock their doors or take the keys out of their cars, the Mugwump shed, and *Rebecca,* are officially locked up. The situation has gotten that bad.

Rebecca, in contrast, stands stately and graceful above it all in the darkened shed, dwarfing these men who argue below her, and dwarfing their affairs. The more devout of the wooden boat worshipers and dreamers might suggest that *Rebecca* herself, knowing that Dan is not her proper owner, has used whatever mysterious power she possesses to bungle his finances. She no doubt is content to wait out bankruptcy court proceedings so that a caring and competent owner may someday be hers.

Nat never fails to maintain that a boat has her own destiny and will one day find her proper owner. In this, as always, he is way up on the spreaders, taking the long view. And more, he knows where the right view is, knows what to set his sights on. He once told me that the situation with Dan didn't bother him because he wasn't building *Rebecca* for Dan, and never had been: "I'm building it for the boat," he said. A simple-sounding but in fact a revealing statement: *I'm building the boat for the boat.*

This is all I will know of *Rebecca,* alas, until some months, maybe years, hence, when she will launch. She *will* one day launch, and she will be the great schooner she is meant to be, and for the proper owner. Nat will make sure of this. But for now she is a statue in a locked room.

II

Elisa Lee never broke stride from the day Ross Gannon finished laying down the lines. Once the boat was rolled, Ross and crew hung the last planks on the raised section of the hull, fastened in stringers and sheer clamps, and began to pound in the deck beams. Ross put in the bulkheads and added more timbers where the engine would rest, and then the big yellow John Deere 225-horsepower was lowered into place, and a plywood deck was fastened down.

Then the main work became a matter of interior carpentry and the insertion of conventional comforts into odd-shaped volumes. Jonathan Edwards returned to the Vineyard with his girlfriend, Cece Hunter, and spent his summer days hanging out in the sun and leafing through a Port Supply catalog, picking out various fixtures and appliances such as a stove and a refrigerator. Ginny grumbled that G&B had never put mechanical refrigeration into a new boat, let alone a house-style refrigerator. Jonathan and Ross conferred daily on the placement of all that *stuff* and the layout of the interior cabin. Brad shook his head and chuckled when he told me that he had to find a spot for an outlet for the electric hair dryer. Jonathan will also ask for a swim platform, another first for G&B.

The work didn't let up in summertime. It intensified through July, partly because it was just a busy time of year—the island at capacity, traffic backed up from Five Corners all the way to Cronig's to the west and the bridge to the east—and partly because everyone seemed to be abandoning the shop. Brad left with his fiancée, April, to sail north and celebrate his fiftieth birthday. Ginny trained her replacement, and then she and Jim bolted for Hawaii to crew on her son's boat on a transpacific voyage. Duane was long gone, after selling his boat, and now spent his days building his own motorsailer in a backyard in Edgartown. Nat left suddenly for a two-week-long religious retreat out west. And Ross blew up at Bob Osleeb, who was spending too much time working on his own boat and taking too long to repair the stem of one of the yard's myriad projects. A brief dialog ensued:

Bob: *"I guess it's time for me to be moving on."*

Ross: *"I guess it is."*

But at least the sun was up early and didn't set till nine, allowing Ross to put in two days for everyone else's one. He set his sights on the first week of August, when he and Kirsten would head to Maine to do some cruising and racing on *When and If* and then participate in the Eggemoggin Reach Regatta. It was a glorious weekend of sailing followed by a big barbecue on the WoodenBoat estate, a gathering of wooden boat lovers—that strange mix of poor builders and rich yachtsmen—who yearly came together as one in a shared adoration of wood in wind and water against the stunning coast and summer skies. When race and race party were over, Kirsten and Ross would rise at four in the morning on Sunday to return by car and ferry to the Vineyard—there was still much work to do on *Elisa Lee,* whose scheduled launch was the following day.

✳

The boat wouldn't head south toward her ultimate home until the fall, after the finish work on the interior, adjustments to the rudder, and installation and swinging of the compass were done—all the myriad details that would complete her. But by midday on Sunday, August 8, the entire G&B crew was at work. The Bella, named *Isabella,* had launched in June, G&B's twentieth boat, and the temporary enclosure was taken down so that *Elisa Lee* was visible from a vantage point that did not foreshorten her. You could step back and look at her entire profile, and suddenly she looked not fat, but rather long and sleek. She had been painted, red lead on the bottom, white enamel on the topsides, and the raised section of the hull was painted aqua—a jarring color on the Vineyard, but one that would be at home in St. Croix. Ross got to work that rainy Sunday, after the drive back from Maine, repainting the waterline. The red lead was to be painted over aqua, and her name had yet to be painted on the transom. Chris Mullen squatted on the port bow, chiseling out of the angelique toerail a hole into which a bronze chock would eventually be fitted. The deck still had to be Dyneled—covered with a synthetic cloth over which epoxy would be squeegeed, resulting in a durable, low-maintenance deck, typical for smaller G&B boats. Ross also had to finish the metal frames for the hatches in the cockpit sole. Rain continued to pour down onto roof and sand, slowing any step that needed time to dry. Kirsten and Simmy sanded and varnished cockpit and cabin.

On Monday morning Ross waited as long as possible before rousing Lyle and bundling him into the truck to go to the boatyard. They arrived at eight-thirty. Nat was already there; he and Ross conferred about the day's events, and then Ross hustled into more work, this morning making corners for the cabin top. David Stimson planed the toerail into its final form, then began making the angelique handrails for the cabin top. Everyone else was sanding, painting, or varnishing. Gretchen, in her loft, sewed the covers for the boat's cushions. During the night the bad weather had pushed through, leaving behind nothing but high-pressure blue sky and August sunshine. Ross installed the windlass in the bow while Nat attached the skylights he'd made of varnished teak, perfectly joined. Brad stood at the helm, head bent down at wires in his hands, connecting the engine controls to the engine.

At one point that morning Nat, returning from the inner shop, paused to look at the boat. She seemed transformed even from yesterday, from rough to polished, a dozen people swarming productively all over her. "Little bit of trim, a little bit of paint, and the whole thing comes alive," he said, grinning.

By midafternoon, with everything done that would be done before the launching, Nat and Ross prepared to roll her out of the shop and over to the new cradle that David Shay had spent all spring welding together. Moving the boat 30 feet east would be simple compared with flipping the thing in a confined space. She was already supported and upright on blocks; they needed only to jack up either end and slide rollers and planks beneath her—two planks, three rollers each. They tied a heavy line to *Elisa* and attached it to the wall to brake any sudden movement. Once she was on rollers and Chris had taken hold of the brake line, everyone else simply put both hands on the starboard side of the hull and pushed, as if moving a great boulder. The boat rolled sluggishly. They stopped and pushed again, and the boat moved more easily and under control. When the back rollers popped out, two people retrieved them, ran them around to the front, and slid them under the planks. Just before she moved over the edge of the floor and onto the sets of planks that served as a ramp over the sand, an enormous crack rang out, and the boat jerked. Ross hustled around front to see what had happened. One of the planks supporting the weight had cracked. Ross said, "We've got a huge roller in here. All the rollers have to be the same size. That's why it broke."

Ross reached for a beam of wood. "I'm just going to use this four-by-four to fulcrum it up." He did, and the proper roller was inserted. Ross sized up the situation again, said to himself, "Is there any reason why we can't come ahead?" Then he said, "Chris, can we have a little more slack?" Half a dozen people pushed the boat onto a declining ramp, and slowly she moved over the cradle. The movement was so smooth and easy that Brad continued to work on the engine controls the whole time. Once she was on the cradle, centered and blocked off, it would be a matter merely of turning on the winch, letting the cable out, and allowing *Elisa Lee* to roll into the water.

Ross walked along the edge of the outer shop, where for six months the boat had lived. The empty shop seemed cavernous and strange now. "Look at that big, empty space—holy mackerel!" he said, happily. "Almost big enough to build a boat in!" He turned to behold the boat on the cradle, the starboard side that had until now been visible only from a few feet away at most. He said, "Haven't seen this side of the boat."

Kirsten, beside him, grinned proudly and said, "She's *gorgeous!* Atta boy." Ross turned to her, smiled shyly, and looked at the sand. "She's all right," he said, with evident pleasure, then headed to the boat to continue readying her for the launch. He turned the last bolt and head on the lathe and inserted them in the back end of the keel, ready to be bunged and painted.

A crowd began to gather when the hour passed 5:00 P.M. A big plastic garbage can, more commonly filled with sawdust and scraps, was lined with a plastic bag and loaded with ice and beer. Jonathan and

Cece set up a table and laid out a half gallon of Mount Gay rum, orange and pineapple juice, plastic cups, and chips and salsa.

Ross's mom appeared with a friend. Ross said, "Hi, Mom. Hi, Ruby." Ross's mom said, "You think you can get this in the water without me?"

"No!" he answered. "That's why it's still here!"

The crowd slowly filling up the shop was a strange mix of folk—Ross stopped every now and then to look around, apparently embarrassed by all the people—of a sort seen only at launchings. While most were friends of the boatyard and like-minded islanders (the rustic earthy fringe, the year-round hippie types and boat people), there were also a number of representatives of the society contingent. One of these last, passing time before the launch, scanned the shop looking for friendly faces and said to me, "These are not my people." Her people, she explained, were the "cocktail-and-tennis set," while these people here were the "Blackfoot tribe." When this woman regarded the new boat, her eyebrows rose, and she said, "It looks like an old boat. It looks like something you'd find washed up on a beach somewhere."

She was right. *Elisa Lee* was not a boat from this era. The beach scene looked like the year 1950 with a fifteen-year-old boat on the cradle. Although brand spanking new—reflecting the images of the boatwrights off her shiny white hull, the planking seams thin but distinctly visible—she was retrograde in design, style, and method of construction. And this fact was evident for the first time only now, when she'd been freed from the shop, was out there, ready to go, ready to become a boat. Brad still worked away on the engine controls. Duane had returned for the launch, and he and Gretchen together placed a small potted palm on the boat's bow. Sawhorses were laid with planks before the stem, where lithe blond Cece would christen the boat. Jonathan draped his arm around her as he addressed the crowd, many of whom had seen him last in concert here on the island.

"Thank you all for coming to our little christening," he said during his brief remarks. He thanked everyone at G&B, thanked Nat and Ross for their work, for their extraordinary craftsmanship, and concluded, "Cece and I really appreciate the lifetime guarantee on parts and labor." Nat laughed loudest.

Jim Martin, the minister at the West Tisbury Congregational

Church, stepped up to offer a blessing. "We're here to celebrate the creation of a vessel," he began. "This is a working boat," he said. "She's built strong by the best, but she's built small, and the ocean is great, so we bless this boat."

Nat hustled through the crowd looking for a stepladder so Cece could climb up on the planks to reach the stem. Both she and Jonathan ascended the makeshift platform. Gretchen appeared with the champagne, concealed within a canvas bag. Cece accepted the bottle, and Nat called out from the crowd, "Right on the bronze!"—meaning, Hit the bottle on that metal stempiece so the bottle's sure to break. Cece delivered a mighty blow, the bottle smashed, the bag tore open, and glass flew. Jonathan and Cece hugged each other while others retrieved thick shards of bottle and the crowd cheered.

A ladder was rested against the lower sheer so that Cece and Jonathan, their dog, a friend of Cece's, Nat, and Kirsten could climb aboard. The boat jerked, and slowly the cable began to unwind. *Elisa Lee* descended toward the water, her crew of five (and a dog) high above the crowd and grinning in the evening sun.

It was an exciting moment. The boat, of course, was supposed to float, but you never knew. It was supposed to float just above the waterline, demarcated in white and aqua enamel. Having helped to position the main keel timber over its drawing on the floor nine months earlier, I found watching this boat at last move toward her proper home, the surface of the water, unspeakably beautiful, exhilarating.

While all eyes were fixed on the boat, on Nat and the others, everyone high with excitement, Ross was way out at the edge of the parking lot, out of view. But it was he who performed the magic of this launch. No one saw him; the wizard worked behind the curtain, while the show went on elsewhere. His right hand gripping the gearshift, in charge of the pawl, Ross gazed calmly at the turning drums of the winch, watching the cable pay out. He appeared to be completely peaceful and at ease.

The boat backed into the water, then floated up off the cradle, 4 or 5 inches of aqua paint visible above the surface. Perfect. You never knew—it was just theory until it floated. But now it was a boat.

Water began to pour into her hull immediately, and Nat hooked up a pump. He'd been expecting this; the wood had dried out considerably after planking. But no one was running into the shop. That was when you could be concerned, when you saw Nat or Ross sprinting for a box of sawdust (dumped into the water near a leak, sawdust helps to clog openings). In an hour the leaking slowed; soon the planks would swell completely shut, and the boat would be sealed.

Visitors were invited aboard as Ross and Brad cranked the engine. Brad, surprised, said, "This has got some *power*." Tora Johnson interviewed Jonathan for her column in the *Vineyard Times* while the crowd milled and devoured pizza. Bob Osleeb had returned to the yard to see the launch. Bob, whose own wooden boat had found him in a storm long ago in Key West, who had done little else since then but work on and sail wooden boats, had gotten a job as a house carpenter since his blowup with Ross. Today was his first day on the job—three stories up in the baking sun, shingling a roof. Why am I doing this? he'd thought. I asked him how it would be, this new work. There, his feet in the sand beside the boatshop, he said, "It's *so* boring." Then he looked out at *Elisa Lee* floating beside the dock, and he pointed to her, a boat he'd been instrumental in building. He grinned and said, "There's *nothing* like *this*."

III

I knew Bob was right—I could feel it, just looking at the boat in the water—but I would understand, eventually, after my family and I packed up and left the Vineyard, that Bob was even more right than he meant at the time. His statement was a worthy comment on the plank-on-frame boat generally—there wasn't anything like it. When you sailed one, it felt substantial, solid, satisfying. And for me it was more than just the fact that sailing a wooden vessel pleased the senses, which it did. What was satisfying to me about moving through water in a planked vessel was the sense of how right it felt. Right in the sense of its being part of a natural order: *This is the way things should work.* The material, wood, and many of its attributes—notably its ability to absorb water and to bend in specific, limited ways—were perfect for creating the shape of a vessel, a shape that happened to move through water in a particularly pleasing way. The pleasure was tactile, like wearing favorite old clothes that had learned the fit of your body. Also, these boats were so damned pretty to look at. Sailing in a fiberglass boat could be pleasurable, too, if you were whipping through the windward leg in a nice light racing vessel, say, or goofing around in a Sunfish; fiberglass did what you wanted it to do, became what you asked, whether fast or cheap. But its pleasures, as far as I had ever known them, had no depth and disappeared shortly after they were experienced.

Before I went to Maine in search of a boatyard, I wrote about food and cooking, and the difference between the two pleasures that I'm trying to convey is to me the difference between eating processed food and food made from scratch. Like eating a Pringles potato chip versus eating a chip that you've sliced off a big russet, fried in hot fat, and sprinkled with coarse salt. I love Pringles; sometimes they're just what I want. This odd creation, chips in a can, stacked—uniform, every one exactly alike, the workmanship of certainty! They're crispy and salty and have almost no flavor. I'll eat them compulsively until they're gone and then feel full, feel nothing for them. I've made some chips for myself that I think I will always remember because they were so satisfying and delicious. But what's important to me is not that one is better than the other; rather, it's that they're completely different things, even though they're both called potato chips. They shouldn't be compared.

Fiberglass boat, cold-molded boat, plank-on-frame boat—they're all different creatures. A plank-on-frame boat is a more complex object, and things that have complexity and also a simplicity of design—as Nat's boats invariably do—are just more pleasing to some people's minds and spirits. Some prefer Lichtenstein; others are transported by Vermeer.

It was this complexity of construction that made the wooden boat so satisfying. Ross had told me he liked wooden boats because you could see all the pieces. The sight of them both engaged the mind and pleased the eye—what more could you want from an object, from a work of art? The wooden boat was a work of art. But it wasn't just some objet d'art—you could sail away on it. Its beauty was secondary. You could sail to Tortola or France or the Galápagos on it. You could do it right now. You could step onto, say, *Liberty*—tugging at her mooring just off the end of the G&B dock—and if you wanted to and had the skill and navigational wherewithal, you could go to Singapore, wouldn't even need much cash. A sturdy boat was freedom. That the beauty of it was so intermingled with its function gave it a depth of pleasure that to me felt bottomless—wooden boats were "forever interesting," as Ross would say.

What was inseparably bound up in my reverence for good wooden boats was my respect for the people who made them. They lived their work. There was an economy and immediacy to their lives. They lived on little, they wasted little, and they had an almost total disregard for the material possessions that increasingly define American culture. They needed wood and they needed boats—after those they could get by on next to nothing.

I'd never before seen a group of people whose ethics were so bound up in what they created. They *were* their boats. There was extraordinary integrity to these vessels—meaning, for instance, that the joints didn't come apart, they were tight and properly made according to the grain of the wood, and the material was the best there was; and meaning also that there was a rightness to bending wood this way and fastening it together and sailing it. So just as there was integrity to these boats, there was a parallel integrity in these boatwrights' lives. For Nat and Ross and Brad, and the G&B crew, their lives were the same as their work, and their work was the same as their lives, the same as the boats. You couldn't say where the life stopped and where the boat began.

Discovering this was urgently important to me because this work and this kind of person were vanishing from our midst. We liked to say we valued those things we called virtues—no matter how old-fashioned and ultraconservative the word might sound—qualities such as self-reliance and resourcefulness. We liked the idea of them, but we didn't do much to foster them in ourselves, nor did we reward others who had them. Self-reliance and resourcefulness, we had none. We were increasingly dependent on electronics, even though we couldn't fix them and didn't understand how they worked in the first place. If our vehicle broke down, few of us knew how to repair it. Call AAA on the cell phone. We wanted services and goods, and lots of them. We were soft. It wasn't long ago that people knew how to do everything as a matter of course. They knew how to build houses. They knew how to grow food crops. How to make cheese, or a bucket or a shirt. They were tough, and they valued the aforementioned virtues because those virtues, like all the other pre–Industrial Age skills, helped them to stay alive and helped them to prosper. We didn't need those qualities anymore in order to thrive. We had to be smart and savvy and cunning today, and we had to work long hours and make our own right-place-at-the-right-time luck, but honesty, self-reliance, and resourcefulness were no longer the assets they'd once been.

And yet these virtues very much helped the wooden boat builders to thrive, and in their work these men reminded us, as McCullough, the historian, said, "how much we're losing in this homogenized, marketing-ethic, throwaway culture."

David Pye and his notions concerning the workmanship of risk—handmade things—illuminated especially the builder of wooden boats, while also pointing to what we stood to lose if we didn't pay attention. It wasn't the plank-on-frame boat, the hand-knitted sweater, the hand-carved dining table, or the Shaker box. There would always be a few who were compelled to create such things, and a few who would buy them. (Or give the worker grants to keep him going: Ralph Stanley, a respected boatbuilder in Maine, received a $10,000 folk art fellowship from the NEA in 1999, as did a basket weaver, a horsehair hitcher, and a tabla player, among others. Wooden boat building was now considered a folk art.) The danger was not that they would cease to be built, but rather that they would cease to be built *well,* and that we would therefore lose our understanding of what was good and what was inferior. That inferior standard would then become accepted as the norm. And once inferior workmanship became the norm, the value of the thing itself would diminish beyond usefulness, and then beyond anyone's memory.

We would enter, or were perhaps entering, a Dark Ages of material things at a time when we increasingly defined ourselves by them. We mass-produced our stuff and made it cheap so we could afford lots of it, and thus the quality had diminished to the point that we needed even more stuff, because cheap things break or go bad or get old fast. Things that grow more valuable with age are typically things of superior quality. Even now we might be unable to recognize things of superlative quality, and if so, it was already too late—we'd have to wait four hundred years for a second Renaissance to flood the culture with light.

You could dismiss this as fancy-pants posturing, pious and insufferable agonizing on behalf of the Truth of the Wooden Boat. I wouldn't say a word. Maybe I didn't know what I was talking about. But I felt pretty certain that, as McCullough had suggested, Nat and Ross *did* know. This fact was underscored for me in mid-October 1999, shortly before *Elisa Lee* headed south toward her home, when I joined Ross, Lyle, their friend Simmy, and Betty, a hyperactive mutt (a stray Ross had taken in), to deliver to G&B an old wooden boat currently tied up in Greenport, Long Island.

Boat deliveries are, as we've seen, invariably dicey, especially when the boat is an old wooden one, as this was: *Jane Dore IV,* a 45-foot sloop. But we knew the form of the diceyness at the outset—the current owner had not had time to fix the steering, so we would use the emergency tiller. The engine had been taken out and was now lashed into the cockpit. But she would be fine on the short sail from the eastern tip of Long Island past Block Island and over to the Vineyard.

We arrived in Greenport after dark, slept on the boat, and woke at dawn the next day, a warm, gray morning. After coffee and some fruit we readied the boat for what would be an all-day sail if the winds were good. Ross maneuvered the boat out of the tricky harbor, set just outside a breakwater. Once we were out of the harbor, the wind picked up quickly, the seas developed a healthy chop, and signs were excellent that it would be a quick, if wet, sail home; we might even arrive before midnight. Ross spotted a man motoring toward us in a rubber skiff—the seller, David Karamidjian, wanting to see us off. David, who had a long blond ponytail, approached the boat, grabbed hold, tied the skiff to *Jane Dore*'s stern, and hopped aboard. Ross and he talked. I took the tiller, and Ross told me to point her closer on the wind, a solid wind at this point. I pulled on the short metal tiller, but the rudder was very hard to move this way; it didn't feel right—it felt like something would give. When Ross saw that I wasn't taking the boat where he wanted it to go, he reached down to help me pull the tiller. I let go, Ross pulled hard, and the tiller broke clean off.

Ross looked at the tiller in his hand for a moment, halted by the surprise of it. The boat was now without steering and had already begun to fly with the current, which led straight into the rocky breakwater. In moments Ross assessed where we were, noted the approaching breakwater, and turned to the seller. *"Do you know how to get the anchor down?!"* he shouted. *"Get it down!"* David ran for the anchor at the bow and threw it over, taking the chain off the windlass drum. Ross hustled to get down the jib, which was pushing the boat off the wind and toward the breakwater. The breakwater seemed to be cruising straight at us. Ross stood over the anchor chain as it rattled like machine-gun fire out of the boat because we were moving so rapidly toward the rocks. He began to step on the chain, trying to slow it. He had no idea how it was tied off at the bitter end, or whether it was tied off at all. The way the anchor was flying, it could rip right out of the boat, and we'd smash into the breakwater in moments. He hammered his foot down onto the speeding chain, then stepped on it with both feet. The chain kept flying out of the boat. He couldn't risk it, had to try: he grabbed for the chain with his big, knotty hands. It ripped right through them until, slowed by Ross's efforts, it reached the bitter end. The boat stopped 150 feet from the breakwater. Another sixty seconds and the boat would have been on the rocks.

We were quiet for a minute. Ross sat in the cockpit. His hands were burned and torn. They quavered involuntarily as he stared at them—not copious blood, but plenty of newly exposed flesh and flaps of skin. The only first aid Simmy could find was an old roll of electrical tape. Ross taped together and covered what he could with that. Then he looked at the "emergency" tiller.

"*Aluminum*," he said. Its center was rotted to powder. "I just *assumed* it was steel." He shook his head in disbelief—an aluminum emergency tiller, several coats of paint obscuring the actual material. Hard to believe, but there it was.

A huge waste of time for Ross (hours more, an entire day, would be spent on buses and ferries with Betty the mutt), burned, cut hands, and a wooden boat that nearly smashed on a breakwater—all because of a bad tiller. Ross could have told you that a tiller was something you wanted to be able to depend on.

✴

The following week *Elisa Lee* was ready to depart the G&B dock for her trip south to the Virgin Islands. Jonathan Edwards was performing in New York and so Ross, still needing to deliver *Jane Dore* from Greenport to the Vineyard, decided to drive *Elisa* to Greenport and sail back, and let Jonathan pick up his boat in Greenport and carry on south from there. The day before he was to leave, Ross had lunch with Nat. They remained unlikely but close friends who laughed a lot together. (Ross: "I tried to get on the Internet for forty-five minutes the other night. Never got on." Nat: "You were low on kerosene— you needed to add some kerosene." Ross: "It's a peat burner!") Ross, looking for company, asked Nat if he wanted to come along to Green-port. Nat said that sounded like a fine idea. So Nat Benjamin and Ross Gannon, alone on a bright, flat-calm morning, would be the ones to steer *Elisa Lee* out of Vineyard Haven Harbor on the first leg of her first big journey, Ross having made sure *Jane Dore* had a good strong emergency tiller waiting for them.

This was the thing about wooden boats. They taught you what was important. Maybe Jon Wilson—now entrenched in a magazine called *Hope*—was right in his grand claim that wooden boats could teach us about our purpose on the planet.

A fiberglass boat can be great or pitiful, but it cannot reach the sorry depths that a neglected wooden boat can. It's only fiberglass, after all; if it's falling apart on its mooring, just throw it away and make a better one—no great loss. An abandoned wooden boat, in contrast, can be an emblem of shame and sorrow and waste—of trees, of lives. It tells us what matters: trees, our lives. And conversely, a wooden boat can come close to glory itself, a practical monument to the natural world, a work of art that bespeaks man's joy in his labor, a functional emblem of self-reliance and adventure. The natural world, art, work, self-reliance, adventure—these things matter.

Ultimately, wooden boats make sense, and have for most of human civilization. They encourage their owners to be the opposite of boat-struck: boat smart. Wooden boats are instructive, each one in its own way. And the wooden boat collectively, as Jon and Nat and Ross all knew at their core, is an important part of our culture, our world, one that should not be lost, especially now, at what may be a critical time in our history. The writer James Salter may have best identified a certain cultural climate of our day, a climate in which wooden boats happen to have begun their small renaissance: "If civilizations reach a new zenith or if they founder is a concern only to us," he wrote, "and not really much of a concern since individually we can do so little about it.

"At the same time, it is frightening to think of a glib, soulless, pop culture world. There is the urge toward things that are not meaning-less, that will not vanish completely without leaving the slightest rip-ple. The corollary to this is the desire to be connected to the life that has gone before, to stand in the ancient places . . . to die in the pres-ence of great things."

I'm not afraid to claim that the wooden boat is both ancient and great, that it connects us to the life that has gone before, and that it's fully worthy of a life engaged in its construction. But Nat and Ross typically stayed away from such talk. They were boatbuilders and sailors—practical men. And so as Nat took *Elisa* out into Vineyard Sound, around West Chop toward New York, on a perfect calm morning, he noted how the shape he had drawn at his board performed, how this hull—which he and his partner, Ross, and Jim Bresson, and Bob Osleeb, and Bruce Davies, and Ted Okie, and Myles Thurlow had all set their hands to, cutting and fastening and fairing and bunging and caulking and sanding—moved through coastal Atlantic waters. He was pleased by her speed and handling. He considered how she steered, how she moved, was she the best possible weight, did that raised deck result in the dry ride he'd predicted? Ross and he talked easily, laughed a lot, enjoyed the ride on a sparkling day, all the while being carried across the water on a vessel that was in fact a series of decisions and woodworking skills, the cumulative effort of a boatyard. The whole G&B crew had had a hand in *Elisa Lee*—Duane and Chris and David Shay and David Stimson and Robert, and Ginny, scavenging parts for the boat and watching her come together before her eyes. Hard-nosed Ginny, wooden boat angel—she owned some of the choicest land on the island, worth millions, but she preferred the work of a wooden boat bookkeeper and pristine, fallow land. And Gretchen, and Kirsten and Simmy, they, too, were in this boat, and Kerry, who arrived from a life of farming to become a wooden boat builder and found that the work suited him. More than a dozen lives, and now the boat was off— this was the end of her beginning. Jonathan Edwards would find her safely tied up in Greenport, and he and Cece would take *Elisa* carefully south from Long Island, Jonathan running her aground only once—no surprise to Nat, *you know how those musicians are*—outside Savannah, Georgia, where she halted harmlessly on sand for all to see; an embarrassed Jonathan stayed below until the tide came in to carry them beyond that single mistake, down to West Palm Beach and then across to St. Croix, where *Elisa Lee* now floats at a dock at the Green Cay Marina, happily soaking up the preserving salt water, bearing the Caribbean sun, a magnet for all who enter that harbor. A new wooden boat has entered the world.

Epilogue

R oss and Nat sailed *Jane Dore IV* uneventfully back to Vineyard Haven and hauled her out. She'd been bought by two
couples from Great Britain, Tim and Josephine Blackman
and Brian and Pam Malcolm, and renamed *Josephine.* The group intended to sail her back to England the following year after numerous
repairs were made by G&B.

During the summer while *Josephine* was hauled, the Malcolms
spent time on the Vineyard. They expressed interest in the schooner
Rebecca. It was in the process of being put up for sale by the trustee in
charge of Dan Adams's bankruptcy filing. Nat and Ross took the Malcolms out for some sailing on *When and If.* The Malcolms decided they
loved the idea of owning a big schooner like *Rebecca,* but they were returning to Scotland and wouldn't be able to do anything with her for
a year. Nat and Ross said they'd be happy to look after her for a year
right there in Vineyard Haven Harbor. After deliberating, the Malcolms
agreed to purchase *Rebecca.* It was done on a handshake. Then Nat
signed a contract to launch the boat by April 30, 2001, and to have her
rigged and sailing by the following June. At the end of October 2000,
the bankruptcy courts approved the sale, and ten days later the sale of
Rebecca closed. The big schooner now has a new owner and an actual
launch date. The Malcolms will retain the name *Rebecca.*

That same October, Nat traveled to Manhattan to meet with
Robert Soros, who had been interested in *Rebecca* but had declined to
make an offer. He wanted his own boat, a little bigger than *Rebecca*
with a little more room below. Nat grinned and said that was a very

sensible decision. The month the sale of *Rebecca* closed, Nat would sign an agreement to build another big boat, a 65-foot schooner, with marconi main and gaff foresail, 52 feet on the waterline, bigger and beamier than *Rebecca*. When the papers were signed, Nat would ask Brad to head down to Suriname to begin looking for timbers, for the big logs, setting in motion the building of yet another wooden boat.

Postscript

Writers love coincidences and tend to make more of them than is useful, but it was personally satisfying nonetheless that *Rebecca* was launched on May 8, 2001, the day after this book was officially published.

I'm happier still to have recently received an e-mail from Ginny Jones noting that on Wednesday, November 28, *Rebecca* sailed into Tortola, having left Bermuda five days earlier ("So she's not only beautiful but a real flyer. . . . Several 200 nm–plus days at the beginning and never turned the key on the engine the whole trip.") It was her first big journey. Ginny also noted that Nat had drawn the lines for *Juno,* the new 65-foot schooner, and that the yard had a total of six boats to build by 2004.

This is good news for G&B, certainly, but it appears to be only part of the larger picture in the world of wooden boats. In May 2001, *WoodenBoat* editor Matt Murphy, describing a tour of Maine boatyards that bustled with activity and new projects, had already noted in his magazine column an apparent renaissance in wooden boat construction. "The number of people employed in wooden boat building and repair has increased dramatically in the past ten years," he told me. "Hard numbers don't exist, but many reputable shops have seen an eight- to tenfold increase in skilled employees in the past decade, and several new shops have opened."

Who knows if this will continue? I hope it does, certainly, but it's idle to guess the future when there are planks to spile, rabbets to chisel, and lines to loft on a newly painted floor.

Michael Ruhlman
January 2002

Acknowledgments
and Sources

My first thanks are to Nat Benjamin and Ross Gannon for letting me poke around in their lives and work. Not only was my presence over many months invasive, it also required a lot of effort on their part. I'm grateful for their trust and their time.

Michael Naumann was the man who suggested I might want to look into the unusual world of wooden boats, and I am grateful that he did.

Ray Roberts is an invaluable editor, as he proved again with this book. Dorothy Straight is an extraordinary copyeditor and gifted reader.

Elizabeth Kaplan, as always, thank you. I wouldn't be here without your counsel and your work.

I'd like to thank all the folks at G&B who were generous with their time, especially those whose place in this book, or lack thereof, inadequately reflects their place at the yard or their help to me, such as Chris Mullen and Gretchen Snyder. The unsinkable Ginny Jones was invaluable throughout my time at the yard and afterward in helping me answer a thousand niggling questions and check countless little facts.

I'd like to thank everyone at *WoodenBoat* magazine, especially Jon Wilson and Matt Murphy. Jon steered me correctly at the beginning and was generous throughout the writing of this book, as was *WoodenBoat* editor Matt Murphy. Matt also sent me to Suriname on behalf of the magazine, a trip that yielded material for this book. Betsy Powell was helpful in tracking down several significant numbers and facts.

Other small but important thanks: Jan Blom offered me a room in Suriname and was helpful with background on the country. Bill Haag gave me a helpful reading of an early draft of the manuscript. Nan Neth gave us a comfortable, affordable home during our months on the Vineyard. Jimmy Carroll provided me with his cell phone when I really needed it. I'm very grateful for the recollections of Tim and Pauline Carr, sent via e-mail from their boat at South Georgia, an island off Antarctica. Beth Gutcheon, thanks for East Egg, where this book began.

I would like to remember two people here, invaluable friends who helped me immeasurably in this work. Rusty King, a great newspaperman and editor, did his best to teach me economy of language, among so many other things. My great-uncle, "Uncle Bill," G. H. Griffiths, pointed me to what matters most. He also owned and sailed wooden boats and was fond of quoting L. Francis Herreshoff, who wrote, "Wood is my sweetheart." I miss them.

Last and most, thank you, Donna, Addison, and James, who come with me wherever I go. I pray you never stop. Donna, thanks for your double work—both at home and behind the Nikon.

✸

While this book is in no way intended to be scholarly I should note a few sources that were particularly helpful. For thematic ideas at the end of Part One and elsewhere, I used *The Book of Wooden Boats,* text by Maynard Bray, and *Wooden Boats from Sculls to Yachts* by Joseph Gribbins, with an introduction by Jon Wilson.

For information on the gaff rig I relied on *Gaff Rig* by John Leather, *Hand Reef and Steer: Traditional Sailing Skills for Classic Boats* by Tom Cunliffe, and *Understanding Rigs and Rigging* by Richard Henderson.

David Stimson's article "The Right Stuff: With Top Materials and Smart Construction, *Rebecca* Is Built for the Ages," in *WoodenBoat,* January/February 1999, was similarly useful.

The article by Vincent Scully "Tomorrow's Ruins Today" appeared in *The New York Times Magazine,* December 5, 1999.

For information on lobsterboats I relied most heavily on *Lobstering and the Maine Coast* by Kenneth R. Martin and Nathan R. Lipfert, as well as *Maine Lobster Boats: Builders and Lobstermen Speak of Their Craft* by Virginia L. Thorndike, and "The Evolution of the Noank Lobsterboat" by Paul Stubing, with Maynard Bray and Meg Maiden, in the March/April 1986 issue of *WoodenBoat*.

Notes pertaining to the history of fiberglass are from "Early Glass: How Fiber, Resin, and Inventive Minds Combined to Create a Materials Revolution" by Stephen Crane, in *Professional BoatBuilder*, December/January 1996.

Katherine Scott wrote a terrific profile of Gannon & Benjamin in the July/August 1990 issue of *WoodenBoat*. I relied on it for historical material as well as for details about the fire.

Information on Park Benjamin is thanks to Bill Benjamin, and some of the family history comes from *Park Benjamin: Poet & Editor* by Merle M. Hoover.

The facts about Malta's bombing in World War II are according to *Webster's New Geographical Dictionary*.

I would never have found the book *The Nature and Art of Workmanship* by David Pye had it not been discussed by Bill Wellman in a *New York Times* Op-Ed article titled "How the Elegant Practice of a Craft Can Be High Tech, Too," December 4, 1999, addressing the question of what defines "high tech" in this Silicon Valley age.

I would like to thank Bob Austin of Williamsburg, Virginia, for writing to *WoodenBoat* with his comments on Sranan Tongo, the creole language of Suriname.

Some details on wood certification were taken from a *New York Times* Op-Ed article, "The Greening of Corporate America" by Jared Diamond, January 8, 2000. Some Suriname information is from Connie Rogers's "Warm Welcome in the Jungle," in the *New York Times* travel section, February 20, 2000.

The quotation from James Salter on page 318 is from "Once Upon a Time, Literature. Now What?," *The New York Times*, September 13, 1999.